Central Advanced Studies

A Dictionary of
PLASMA PHYSICS

Central Advanced Studies

A Dictionary of PLASMA PHYSICS

Edited by

Professor Satyendranath Goswami
MSc (Cal. U), DPhil (Cal. U), PhD (Durham)

New Central Book Agency (P) Ltd
LONDON
HYDERABAD ERNAKULAM BHUBANESWAR
NEW DELHI KOLKATA PUNE GUWAHATI

NCBA

REGD OFFICE
8/1 Chintamoni Das Lane, Kolkata 700 009, India
email: ncbapvtltd@eth.net

OVERSEAS
NCBA (UK) Ltd, 149 Park Avenue North, Northampton, NN3 2HY, UK
email: ncbauk@yahoo.com

EXPORT
NCBA Exports Pvt. Ltd, 212 Shahpur Jat, New Delhi 110 049
email: ncbaexports@ncbapvtltd.com

BRANCHES

212 Shahpur Jat
New Delhi 110 049
email: ncbadel@ncbapvtltd.com

Shop Nos. 3 & 4, Vinayak Towers
681/B Budhwar Peth, Appa Balwant Chowk
Pune 411 002
email: ncbapune@ncbapvtltd.com

Shop No. 15, Bharati Towers, Ground Floor
Forest Park, **Bhubaneswar** 751 001
email: ncbabub@ncbapvtltd.com

House No. 3-1-315, 1st Floor
Nimboliadda, Kachiguda, **Hyderabad** 500 027
email: ncbahydb@ncbapvtltd.com

GSS Shopping Complex, 1st Floor
Opposite College Post Office, Convent Road
Ernakulam 682 035
email: ncbaernk@ncbapvtltd.com

Radhamayee Complex (Second Floor)
Opp. SBI, Panbazar, Jaswanta Road
Guwahati 781 001
email: ncbaguwahati@ncbapvtltd.com

A Dictionary of Plasma Physics • Professor Satyendranath Goswami

First Published: 2006
Reprinted: 2013

PUBLISHER AND TYPESETTER
New Central Book Agency (P) Ltd
8/1 Chintamoni Das Lane, Kolkata 700 009

PRINTER
New Central Book Agency (P) Ltd
Web-Offset Division, Dhulagarh, Sankrail, Howrah

COVER DESIGNER
Soumen Paul

CONSULTANT
Professor Santwana Ray Chaudhury
PhD (Cal. U.), Saha Institute of Physics

COMPUTER APPLICATION
Rabindra Nath Goswami
MSc (K.U.)

ISBN 978 81 7381 516 4

Price: ₹ 150.00

PREFACE

In recent years, the subject of Plasma Physics has been expanding at a rapid rate. Being a late starter, the subject, however, does not have some working dictionaries, the usefulness of which cannot be over-emphasized. This volume is intended to provide a general information on the subject. To a wide range of readers, teachers, and researchers in Plasma Physics, this dictionary is capable of giving the basic knowledge about the subject.

The arrangement of words has been done necessarily alphabetically. There are some illustrative diagrams.

The compilation work and editing are carried out by using the following standard texts on the subject, in addition to the other recent sources, namely, various websites.

1. *Introduction to Plasma Physics and Controlled Fusion*—Francis F. Chen.
2. *Plasma Physics*—B. Samuel Tanenbaum.
3. *The Plasma State*—Juda Leon Shohet.
4. *Physics of High Temperature Plasmas*—George Schmidt.
5. *Introduction to Plasma Physics*—B.M. Smirnov.
6. *Elements of Plasma Physics*—S.N. Goswami.
7. *Plasma, the Fourth State of Matter*—D.A. Frank-Kamenetskii.
8. *Plasma Physics*—E.W. Laing.
9. *Plasma Dynamics*—Boyd and Sanderson.
10. *Theoretical Methods of Plasma Physics*—Van Kampen and Felderhof.
11. *Fundamentals of Plasma Physics*—S.R. Seshadri.

In spite of the care taken to include the latest topics, (which become old soon due to the rapid rate of progress of the subject) omissions and errors are likely to exist abundantly. The editor would welcome comments and suggestions from readers, on the material, so that they can be incorporated in future editions.

S.N. Goswami

A

a.c. conductivity it is defined by $\sigma = n^0e^2/m(\nu - i\omega)$ where m = electron mass, ν = electron collision frequency, e = magnitude of electron charge, ω = wave frequency. For large ν, this approaches zero, so that no current flows in the plasma and the power absorbed is zero.

absolute instability an absolute instability is a disturbance that grows in time and does not decay at any fixed point, but is allowed to move. This means that an observer at any fixed point will see the disturbance begin from zero and then increase without limit.

absolute plasma instabilities a class of plasma instabilities with amplitudes growing with time at a fixed point in the plasma medium.

absorption line in spectroscopy, a characteristic wavelength of emitted radiation that is partially absorbed by the medium between the source and the observer.

absorption of plasma wave energy the loss of plasma wave energy to the plasma particle medium. For instance, an electromagnetic wave propagating through a plasma medium will increase the motion of electrons due to electromagnetic forces. As the electrons make collisions with other particles, net energy will be absorbed from the wave.

acceleration the speeding up of motion(or any change of velocity, in magnitude and/or direction).Fast electrons in the aurora, charged particles in the radiation belt, cosmic rays etc., all require an acceleration process to provide their high energy.

acceleration of cosmic rays it was proposed by Enrico Fermi (1901-1954) in 1949 that a charged particle trapped between two magnetic mirrors approaching each other can acquire acceleration. This acceleration is known as **Fermi acceleration.**
It is a possible mechanism by which two stellar clouds moving towards each other impart acceleration to a cosmic charged particle.

acceleration of plasma conducting plasmas may be accelerated. That is, if an external current is passed through a plasma in a magnetic field so that it is perpendicular to $\vec{B}$, a body force is set up that can accelerate the plasma (or conducting fluid). The reverse situation is also true. That is, if some external force pushes the plasma through a magnetic field, an electromotive force may be set up that can produce electrical energy. This electromotive force can be used to measure the velocity of the plasma or to produce a significant amount of electric power for external consumption.

accelerator a device which uses electric or magnetic fields to increase the speed of charged particles. In magnetic fusion, neutral beam heating is used on present Tokamaks (q.v.) to heat the plasma to fusion relevant temperatures. These accelerators boost large quantities (tens of amperes incurrent) of positively charged deuterium ions to 80-120kev to be injected into the plasma. Future accelerators for magnetic confinement are expected to boost particle energies to ~500kev studying the make up of the nucleus, such as in high energy physics research, requires accelerators capable of boosting a particle's energy into Mev range. Present-day accelerators can produce particles with energies well above one hundred billion electron volts— allowing them to probe the constituents of nuclei.

acoustic dispersion relation equation of the type ω = group velocity. K, which relates the wave frequency ω to the wave number K, is called the dispersion relation.

acoustic wave this is a sound wave in a mixture of two gases, namely, electron gas and ion gas, at different temperatures. There is no influence of the charges, since electrons and ions slowly oscillate together.

active region a localized transient volume of the solar atmosphere in which sunspots, facula, flares etc. may be observed. Active regions are the result of enhanced magnetic fields; they are at least bipolar and may be complex if the region contains two or more bipolar groups.

additional heating heating additional to Ohmic heating. Used to heat TOKAMAKs (q.v.) to temperatures at which Ohmic heating is small. Usually uses neutral beams or radio-frequency waves. Also called Auxiliary Heating. See electron cyclotron resonant heating, ion cyclotron resonant heating, lower hybrid heating.

adiabatic compression also called **magnetic heating,** in which the energy of the particles is increased by slowly increasing a magnetic field in the plasma. There is no loss or gain of heat from outside. Magnetic moment of plasma particles remains constant. A special case is the **pinch effect.**

adiabatic heating heating additional to Ohmic heating. Used to heat TOKAMAKs to temperatures at which Ohmic heating is small. Usually uses neutral beams or radio frequency waves. Also called Auxiliary Heating.

adiabatic plasma plasma confinement system involving particles where the orbit radius and orbit period are small compared to the characteristic scales of length and time. Here individual particles closely follow the magnetic field lines by tightly circling them.

advanced TOKAMAKs TOKAMAKs are naturally pulsed devices because the plasma current is driven by inductive means (by a transformer). However, it is possible that so-called "Advanced TOKAMAKs" are feasible; these would operate continuously with a current driven by a combination of non-inductive external drive and the natural pressure-driven currents that occur in plasmas. They would require careful optimization of pressure and confinement. They are being studied both theoretically and experimentally (at Culham, JET and elsewhere) as continuous operation is highly desirable for fusion power production and their relatively small size results in a more economical power plant than an ITER-like design.

ae index a geomagnetic index describing the auroral electro jet.

Alfven gap modes the toroidal nature of TOKAMAK plasmas produces gaps in the otherwise continuous spectrum of Alfvén waves which are populated by discrete, undamped Alfvén gap modes. These modes could be easily destabilized by resonant energy transfer from energetic particles (e.g. alpha particles from fusion reactions).

Alfvén ion cyclotron instability an electromagnetic micro-instability near the ion cyclotron frequency, driven by the ion loss cone in a mirror device.

Alfvén resonant layer the Alfvén resonance layer can be located at any desired position between the plasma centre and the plasma edge. It depends on the appropriate choice of frequency and mode numbers.

Alfvén time the time taken by an Alfvén wave to travel one radian in the toroidal direction. This is a measure of the time-scale on which Alfvén MHD effects can occur.

Alfvén velocity it is given by $V_A = \sqrt{B^2/\mu_0\sigma}$, where B = externally applied magnetic field, σ = mass density of the plasma, and μ_0 = permeability of vacuum.

Alfvén waves waves moving with Alfvén velocity. They are also called hydromagnetic waves or magneto-hydrodynamic waves. First predicted in 1942 by Hannes Olof Gosta Alfvén (1908-1996)

alpha particle (1) A type of fast ion emitted by many heavy radioactive nuclei, such as uranium. Actually, the nucleus (atom stripped of all electrons) of the gas helium.

(2) He_2^+, a positively charged particle consisting of 2 protons and 2 neutrons, denoted by the Greek letter, alpha(α), a helium-4 nucleus. A α-particle is a typical product of fusion reactions.

(3) The nucleus of helium atom, consisting of two protons and two neutrons bound together in a fusion power plant, energetic α-particles (as well as neutrons) will be created by the fusing of deuterium and tritium nuclei. The heating which is provided by these α-particles as they slow down due to collisions will be essential in achieving ignition.

ambipolar diffusion it is a combined diffusion of oppositely charged particles, namely ions and electrons, so that the two fluxes are equal. The coefficient is given by $D_a = (\mu^+ D^- + \mu^- D^+)/(\mu^+ + \mu^-)$ with usual meaning of symbols.

ambipolar plasma potential electric fields that are self-generated by the plasma and act to preserve charge neutrality through ambipolar diffusion.

ampere's law in mathematical form, it is $\int \vec{B}.d\vec{e} = \mu_0 I$, where the symbols have usual meanings.

anisotropic pressure phenomena, in which the properties of a medium are different in different directions, are called anisotropic. The longitudinal temperature T_{11} and the transverse temperature $T\perp$— each of these temperatures can be used to define a pressure in its direction:
$P_{11} \neq P_1$. Therefore, tenuous (or thin) plasmas located in a magnetic field exhibit an **anisotropic pressure.** As the density increases, energy exchange between the various degrees of freedom is facilitated and the anisotropy of the temperature and the pressure is equalized.

anomalous diffusion a process in which the diffusion of plasma perpendicular to the magnetic lines of force occurs much more rapidly compared to the diffusion rate determined by the classical formula in the magnetic field.

anomalous dispersion as long as the oscillation frequency is small in comparison with the cyclotron frequencies, the propagation velocity does not depend on the frequency, i.e. there is no dispersion. In this range, the phase and group velocities coincide. At frequencies comparable to the cyclotron frequencies, however, dispersion effects appear, i.e., the velocity of propagation starts to depend on the frequency. In this frequency range, the phase velocity is not the same as the group velocity. Especially strong dispersion is observed near the two characteristic frequencies at which the refractive index of the plasma tends to infinity, i.e., the phase velocity reduces to zero. The anomalous phenomenon in optics is called **anomalous dispersion.**

anomalous dissipation it is a process peculiar to plasmas and occurs as a result of the long-range interaction of particles with their self-excited fields. In fully ionized hot plasma, collisions are rare and the damping is due to anomalous dissipation. In the simplest kind of oscillation—electrostatic plasma waves—this process is known as **Landau damping.**

anomalous skin effect in plasma, it is possible to find stationary states with a prescribed incident electromagnetic wave. It is found that the single mode that obeys the dispersion law predominates, but there is an admixture of the other modes with the same ω. Consequently, the reflection and transmission coefficients deviate slightly from the values given by Fresnel's formulas. The physical reason is that the electrons in a plasma move freely under the influence of the field, while the electrons in a dielectric medium are elastically bound to their equilibrium positions. This deviation from classical optics is called **anomalous skin effect.**

anomalous transport measured heat loss is anomalously large compared to basic collisional theory of heat transport in toroidal plasmas ("neoclassical" theory). Particularly true for electrons.

aphelion that point on the path of a Sun-orbiting body most distant from the centre of the Sun. Compare **perihelion.**

apogee that point on the path of an Earth-orbiting satellite most distant from the centre of the Earth. Compare **perigee.**

Appleton-Hartree dispersion relation this is used to study the propagation of radio waves through the ionosphere, taking account of the earth's magnetic field. In addition, it is possible to draw, with the help of this equation, the graph of wave phase velocity and frequency for any angle between the steady applied magnetic field and propagation vector.

arc discharge in the V-I characteristics of a gas discharge, there is a region of high current (≈10^3 amp) and low (≈1V) voltage, which is known as the arc discharge.

arc plasma (i) A type of electrical discharge between two electrodes characterised by high current density within the plasma between the electrodes.
(ii) A form of electric discharge which can generate heat with a high energy density of several ten thousand degrees Kelvin in a short period of time

arcade a series of magnetic loops, overlying a solar inversion line.

arch filament system (AFS) a system of small, arched linear-absorption features connecting bright, compact plages of opposite polarity. An AFS is a sign of emerging bipolar magnetic flux and possibly rapid or continued growth in an active region.

ARIES a comprehensive TOKAMAK fusion power plant study.

ASDEX (axisymmetric divertor experiment) a TOKAMAK at the Max Planck Institute fur Plasma Physics, Garching, Germany, was designed to study the effect of a double null divertor.

aspect ratio the ratio of major to minor axis of a torus.

associative lonization this can be represented by the symbolic equation:

$A + B^* \rightarrow AB^+ + e^-$.

astron it is a thermonuclear machine based on confinement of ions by an E layer. Energetic (relativistic) electrons are injected into a magnetic field terminated by magnetic mirrors to form an E layer (a thin solenoidal layer) creating a magnetic bottle of closed field lines with the ions confined in the bottle.

astronomical unit (AU) the mean distance between the Earth and Sun, equal to 214.94 solar radii or 1.496E + 11m Atmosphere. The gaseous envelope surrounding a planet or star.

(1) The mean distance between the Earth and Sun, equal to 214.94 solar radii.

(2) The mean Earth-Sun distance, a unit of distance widely used in expressing distances in the solar system.

astrophysical plasmas they include the sun and other and other stars, the solar wind stellar winds, large parts of the interstellar medium and the intergalactic medium, nebulas and more planets, neutron stars, black holes and some neutral hydrogen clouds which are not in a plasma state. Approximately 99% of the observable universe can be described as being in a plasma state.

astrophysics the branch of astronomy and physics that deals with the physics of astronomical objects.

atmosphere (1) The gaseous envelope surrounding a planet or star.

(2) The layer of gas surrounding the earth or other planets. The upper atmosphere is the region of Earth's atmosphere above the troposphere. Regions of the upper atmosphere are the stratosphere, mesosphere and thermosphere.

EARTHS ATMOSPHERE

No	Layer	From (km)	To (km)	Thickness (km)
1.	Troposphere	0 (Earth)	10	10
2.	Stratosphere	10	40	30
3.	Mesosphere	40	80	40
4.	Thermosphere	80	700	620
5.	Exosphere	700 Interplanetory —Space		—

atmosphere of other planets different types of ions are likely to exist in the atmosphere of other planets than the earth. In the atmosphere of planets Venus and Mars, CO_2 is known to exist. In Venus, HCO_2^+ is produced at a high rate. In the atmospheres of other distant planets, plenty of H_2 and, in their ionospheres, H_3^+ have been found to exist. Compared to the atmosphere of the earth, that of Mars is of lower density, and so binary collision and radiative association are the principal recombination processes occuring there.

Information received from Jupiter and Mercury are numerous. Particularly the following reaction has been proposed for

the structure of the ionosphere of Jupiter:
$H_e^+ + H_2 \rightarrow H_eH^+ + H \rightarrow He + H_2^+$
Space probes have recently supplied much information about upper atmospheres of the other planets of the solar system.

atomic energy energy derived from the mass-to-energy conversion ($m = \frac{E}{C^2}$) that occurs as a result of nuclear reaction.

atomic mass unit (AMU) a unit of mass equal to exactly 1/12 the mass of a carbon atom of mass number 12.1; amu = 931.48 MeV = 1.661×10^{-27} kg.

atomic number (Z) the number of protons in the nucleus of an atom.

attachment

dissociative, $AB + e^- \rightarrow A^- + B$

molecular, $A_2 + e- \rightarrow A_2^-$

radiative, $A + e^- \rightarrow A^- + h\nu$

attitude (of a satellite) the direction in which the satellite is oriented in space.

aurora (australis, borealis) it is a natural display of light in the polar sky and can be seen with the unaided eye only at night. An auroral display in the southern hemisphere is called aurora australis or southern lights. A similar phenomenon in the northern hemisphere is called the aurora borealis or the northern lights. Auroras are the most visible effect of the sun's activity on the earth's atmosphere. Auroras appear chiefly as arcs, clouds and streaks. Some auroras move, become brighter or flicker suddenly. The most common colour in an aurora is green. Most of them occur about 100 to 1,000 kilometers above the earth. Auroral displays are associated with the solar wind, a continuous flow of electrically charged particles from the sun.

auto-ionization it is a process in which a negative molecular ion has a finite life-time and can decay into a molecule and a free electron, It can be represented by: $A^- \rightarrow A + e^-$.

avalanche breakdown plasma can be created in a semi-conductor by applying an external field strong enough to cause this breakdown. This process occurs when the electrons gain enough energy from the applied electric field to dislodge other electrons when they collide with lattice electrons, so that a high electron number density is quickly created.

B

B.B.G.K.Y. equations the $(\partial f/\partial t)$coll term of the Boltzmann equation is quite complex for calculation. To calculate this term, an approach, different from the expansion of Boltzmann's collision integral, was adopted or developed independently by Bogoliubov, Born and Green, Kirkwood, and Yvon. This is why the equations obtained by the method are known as the B.B.G.K.Y. equations. Here, initially, the particle interactions are treated exactly. Needless to mention, this exact treatment of the interactions leads to complexities of other kinds that make it virtually impossible to apply the approach to most of the practical problems that have been solved using the Boltzmann equation. In spite of this, the method has been applied to some problems and can also be used for clarifying the approximations involved in using the BV equation or any of the more detailed collision models. One should remember that almost any analysis using these equations becomes extremely involved.

Balescu-Lenard equation it is another kinetic equation for plasmas, like B.B.G.K.Y. equation. It is given by $\partial f_1/\partial t = -\partial/\partial\vec{V}_1.J(\vec{V}_1)$. The left hand side is $\partial f_1/\partial t$ and the right hand side remains only a function of $\vec{V}_1$ after all of the integration have been performed. The method for obtaining this equation is to assume that f_1 is constant, i.e. $\partial f_1/\partial t = 0$ over the time of integration, as suggested by Dupree.

"Ball" lightning if no confinement surface using currents and magnetic fields can exist in a completely closed spherical geometry, then ball lightning must be confined by another process, which is perhaps only atmospheric pressure.

ballooning instability a local instability which can develop in the TOKAMAK when the plasma pressure exceeds a critical value, it, therefore, constrains the maximum β that can be achieved. It is analogous to the unstable bulge which develops on an over-inflated inner tube.

ballooning mode a plasma mode, which is localized in regions of unfavorable magnetic field curvature (also known as "bad curvature") that becomes unstable (grows in magnitude) when the force due to plasma pressure gradients is greater than the mean magnetic pressure force.

banana orbit the intersections of magnetic surfaces with a plane, containing the axis of symmetry, are nested loops, represented by the concentric circles of Fig. (see p.394, Schmidt). To the lowest order in the guiding centre approximation, a particle would follow a field line, with the guiding centre sliding on a magnetic surface. To the next order, the $\vec{\nabla}$ B and centrifugal drifts emerge, carrying the particle, say upwards. It is noted that it is just this drift that prohibits particle confinement in a toroidal field without a poloidal component. In the case under consideration, the drift will carry the particle to inner magnetic surfaces in the lower half and to outer surface in the upper half of the figure. The resulting may map onto a line on the cross-section; hence the particle is confined.

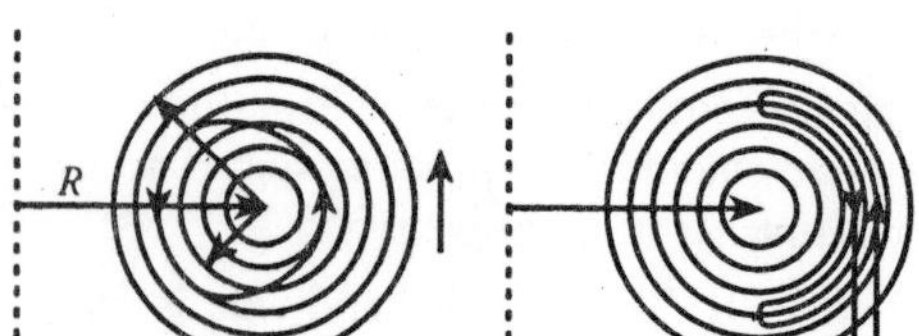

(a) (b)

Most particles, however, behave quite differently. The field is stronger closer to the axis than further way, forming magnetic mirrors. Consequently, many particles are prevented from going around the toroid the short way and are trapped. Because of its shape, this has been named **banana orbit**. The majority of the particles in a tokamak are trapped.

b-angle as viewed from the earth, the heliographic latitude of the centre of the solar disk. The centre of the solar disk usually does not coincide with the heliographic equator, due to a tilt of the solar axis with respect to the ecliptic.

barometric formula the distribution of molecules in the earth's atmosphere by height under the action of the gravitational field is given by

$N = N_0 \exp(- Mgh/ T)$, where

M = molecular mass, g = acceleration due to gravity, h is the height above the earth's surface, N(0) = particle density at the origin, N = particle density at height h, T = temperature in the absolute scale.

baseball coils coils (copper or superconducting) that carry electric current for producing magnetic fields that are shaped like the seams of a baseball, also known as yin-yang coils.

beam-plasma instability this is an important example of instability, the mechanism of which is as follows: suppose that an electron-beam penetrates a plasma and that the velocity of the electrons in the beam is much higher than the thermal velocity of the plasma electrons, while the density of the electrons in the beam is much lower than the density of the plasma electrons. Deceleration of the electron beam can occur owing to the scattering of beam electrons on the plasma electrons and ions. Plasma oscillations generated in the plasma interact with the beam electrons and take energy from them and thus get amplified to give rise to instability. The energy of the electron beam is thus transferred into the energy of plasma oscillations and remains in the plasma.

beam plasma reaction fusion reaction that occurs in neutral beam heated plasmas from the collision of two fast ions originating in the neutral beams injected into the plasma for heating purposes. Distinguished from beam plasma, beam-wall, and thermonuclear (plasma-plasma) reactions.

beams of plasma most (99%) of the universe, with a few exceptions like the earth, is in plasma state. First, these are the stars, which because of their high temperatures, are almost completely ionized. Next, there is the interstellar gas, which, despite its low number density, still exhibits some plasma properties. Finally, there are the numerous special plasmas that show up in a wide variety of geophysical and astrophysical problems. These include the ionosphere, the region of partly ionized gas that extends from about 50 km above the earth to several earth radii, the Van Allen belts of energetic ionized particles trapped at high altitudes by the earth's magnetic field; the beam of plasma which erupt from the sun during magnetic storms and which are related to the spectacular aurora borealis seen in the earth's upper polar atmosphere; the steady streams of plasma known as the **solar wind**, which greatly distorts the earth's magnetic field at heights greater than two or three earth radii; and the spectacular gas discharges created by a lightning stroke.

Beltrami Fields. Beltrami Flows the special case in which the current and magnetic field are parallel, resulting in the vanishing of the electromagnetic body force; is of interest in certain astrophysical situations. Such magnetic field configurations are said to be **force-free**. The problem arose originally from a situation in which a very tenuous ionized gas carried a strong magnetic field, so that the gradient of the fluid pressure was at least an order of magnitude smaller than the magnetic pressure gradient. Then we have

$$\vec{\nabla}(B^2/8\pi) \approx (\vec{B}.\vec{\nabla})\vec{B}/4\pi$$

Or $\vec{B}\times(\vec{\nabla}\times\vec{B}) \approx 0$

There is, then, a parallel with hydrodynamics where velocity fields exist, for which $\vec{U}\times(\vec{\nabla}\times\vec{U}) = 0$.

Such vector fields are known as **Beltrami fields** and flows satisfying the last equation are called **Beltrami flows**. These flows are important in certain practical hydrodynamics and aerodynamic applications.

Bennett Relation in the case of the linear pinch, the Bennett relation $I^2 = 4NKT$ holds. It shows that the plasma temperature T is proportional to the square of the discharge current I in the filament and inversely proportional to the particle line density N.

Bernstein Mode type of plasma mode that propagates perpendicular to the equilibrium magnetic field in plasma. Bernstein waves, named after the plasma physicist Ira Bernstein, have their electric field nearly parallel to the wave propagation vector and their frequency between harmonics of the electron cyclotron frequency.

Bernstein Waves Electrostatic waves propagating at right angles to $\vec{B}0$ at harmonics of the cyclotron frequency are called Bernstein waves.

Bers and Briggs criteria for instabilities bers and Briggs developed a criterion for distinguishing certain types of instabilities. Their criterion states that an absolute instability occurs if two poles, one from the upper half K plane, merge as ω goes from any given value of ωL down the imaginary ω axis in the ω plane. This merging process results in a branch line beginning at the merged poles.

If no poles merge as the frequency is changed, the system may then have only convective or evanescent modes. If poles cross the real K axis as the frequency is moved to the real ω axis, they are convective instabilities. If they do not cross this axis, the waves are propagating.

beta (β) it is the ratio of the plasma pressure p to the magnetic pressure $B_0^2/2\mu_0$ of the confining field.

(1) The ratio of plasma gas pressure to magnetic field pressure in a TOKAMAK.

(2) Ratio of plasma pressure to magnetic field pressure. The highest value of beta achieved in a large TOKAMAK is about 13%, though much higher values are theoretically possible at low aspect ratio and have been achieved on START.

beta (β) for a plasma the quantity β for a plasma $\beta = \dfrac{p}{B_0^2/2\mu_0}$ is defined as the ratio of kinetic pressure to magnetic pressure. In general, the lower the β is, the more likely a plasma is to be confined, if β is computed by taking the kinetic pressure over the magnetic pressure when no plasma is present. β should never be greater than 1 for this type of confinement.

beta limit (1) Maximum beta attainable, usually due to deterioration in the confinement.

(2) Maximum achievable ratio of plasma pressure to magnetic pressure for given plasma to remain stable. In a tokamak, if the beta value is too high, ballooning modes become unstable and lead to a loss of plasma confinement.

beta particle an electron emitted from the nucleus of a radioactive element and denoted by the Greek letter beta.

binding energy in nuclear reactions, the energy associated with removing protons and/or neutrons from the nucleus; in standard chemistry, it is the energy associated with electronic bond making and breaking.

bipolar magnetic region a region of the solar photosphere containing at least two areas of enhanced magnetic fields of opposing polarity.

Birkeland currents Electric currents linking the Earth's ionosphere with more distant regions, flowing along magnetic field lines. Named for Kristian Birkeland (1862-1912), a pioneer of auroral research who first proposed such currents in 1903, these currents are often associated with the polar aurora.

blackbody temperature the temperature of an object if it is reradiating all the thermal energy that has been added to it. If an object is not a blackbody radiator, it will not reradiate all the excess heat and the leftover will go towards increasing its temperature.

blanket (1) The physical system surrounding the hot plasma that provides shielding and absorbs fast neutrons converts the energy into heat and produces tritium.

(2) In a fusion power plant using deuterium-tritium fuel, the system surrounding the plasma vessel slows down the neutrons produced, so that the heat released can be used for electricity generation. In many designs, the blanket is also used to synthesize tritium to use as fuel.

Bohm criterion The condition $KT<m_i u_0^2$ is known as the Bohm criterion for the formation of a plasma sheath (symbols have usual meaning).

Bohm-diffusion it is found that in measuring the diffusion rates of plasma across the magnetic field lines, the rate is often faster than that predicted by the classical theory. The reason for this is thought to be a turbulence or instability developed by the unequal rates of diffusion across the magnetic field.

An empirical formulas developed by Bohm (1946) gives the perpendicular diffusion rate as $D_{Bohm} = 10^{-4}KT/16q_eB$, where K = Boltzmann's constant and T is the plasma temperature. Very often, this diffusion coefficient (and associated diffusion rate) is used for comparison with experimental measurements.

Bohm transport is the anomalous diffusion associated with long wavelength plasma fluctuations which has the consequence that confinement times increase linearly with magnetic field. In gyro-Bohm transport, fluctuations have a shorter scale length, comparable with the ion gyro-radius and, consequently, the confinement time increases quadratically with magnetic field. Gyro-Bohm transport is, therefore, more optimistic than Bohm for large scale devices like ITER.

Bohm-Gross dispersion relation the equation

$\omega^2 = \omega_p^2 + 3K_1^2 < V^2 >$ where $<V^2>$ is the mean square velocity of the distribution(in the case of one dimension) and K_1 is the propagation vector, is known as the Bohm-Gross dispersion relation. This equation is valid for Maxwellian distribution. Its use is suitable in the case when the electrons move only along the direction of wave propagation, that is, when their degree of freedom is one. Another point to remember is that, for the derivation of the equation, all $V > \omega/K_1$ has been disregarded.

Bohm-Sheath criterion the inequality $u_0 > (KT_e/M)^{1/2}$ is known as Bohm-Sheath criterion.

Bohr-Einstein radiation formula the internal electronic energy changes of an atom are connected to the frequency of the corresponding emitted radiation by the formula $E = h\nu$, where h is the Planck's constant. Usually this equation is assumed to determine uniquely the resulting intensity of the radiation. However, there is theoretical and observational evidence that this assumption is only valid if the broadening of the spectral line due to plasma field fluctuations (Stark broadening) is small compared to the natural broadening.

Boltzmann distribution statistical physics proves that in thermodynamic equilibrium (i.e. in a collisionally determined closed system) the volume density of particles decreases exponentially with increasing energy, i.e. $f(E) = \exp(-E/E_0)$.
The energy distribution of electrons within an atom is generally assumed to behave in this way.
However, in more practical cases, collisions are quite insignificant compared to radiative processes which are determined by the lifetime of the individual atomic levels. As a consequence, the distribution function has very little to do with a Boltzmann distribution.

Boltzmann equation the equation is given by

$$\partial f/\partial t + V.\nabla f + F_e/m.\nabla_v = (\partial f/\partial t)_{coll}$$

where the symbols have their usual meaning, $f = f(\vec{r}, \vec{V}, t)$. Boltzmann's equation can be derived from equation of motion. Conversely, the equation of motion can be derived from the Boltzmann's equation. The two principles are equivalent, neither one contains more information than the other. Boltzmann's equation is, however, better for handling collective phenomena.

Boltzmann h-function: h-theorem for a homogeneous gas, Boltzmann's H-theorem states that the quantity

$$H = \int f \ln f \, d\vec{r} d\vec{V}$$

which is known as the Boltzmann H-function obeys the relation

$$\partial H / \partial t \leq 0.$$

The equilibrium condition is determined by $\partial H / \partial t = 0$.

Boltzmann's collision integral the term $(\partial f/dt)_{Coll}$ denotes the rate of change of f due to collisions and its computation is quite complex. Calculation of this term for short-range and long-range forces can be made. In the calculation of binary elastic collisions, an important term known as the **Boltzmann's collision integral** is obtained, which, expanded, gives the **Fokker-Planck** equation. Both of them are of much importance towards understanding of the properties of the plasma.

Boltzmann's transfer equation the fundamental equation used for the macroscopic approach of study of a plasma is known as the Boltzmann's transport or transfer equation. It describes changes in the position and velocity of charged particles due to their own interactions r due to electromagnetic fields, externally applied or internally produced by their motions.

Boltzmann-Vlasov (B.V.) equation: when the collision term $(\partial f/\partial t)_{Coll}$ in the Boltzmann equation is zero, the resulting equation, namely

$$\partial f/\partial t + \vec{V}.\vec{\nabla}_r f + \vec{f}_{e/m}.\vec{\nabla}_v f = 0$$

is called collisionless Boltzmann-Vlasov equation or B-V equation.

bootstrap current (1) Currents driven in toroidal devices by neo-classical processes. They may amount to a substantial fraction of the net current in a TOKAMAK reactor, thus lengthening the pulse time or decreasing the power needed for current drive.

(2) theory predicted in 1970 that a toroidal electric current will flow in a TOKAMAK which is fuelled by energy and particle sources that replace diffusive losses. This diffusion driven bootstrap current, which is proportional to beta and flows even in the absence of an applied voltage, could be used to provide the confining magnetic field : hence the concept of a bootstrap TOKAMAK ~~which~~ has no toroidal voltage. A bootstrap current consistent with theory

was observed many years later on JET and TFTR; it now plays a role in design of experiments and power plants (especially advanced TOKAMAKs).

bounce frequency the average frequency of oscillation of a particle trapped in a magnetic mirror as it bounces back and forth between its "turning points" in regions of high magnetic field.

bow shock (1) a collision shock wave in front of the magnetosphere arising from the interaction of the supersonic solar wind with the Earth's magnetic field.

(2) A collision shock wave in front of a planetary magnetosphere, the place where the supersonic flow of the solar wind is slowed to subsonic speed by the planetary magnetic field.

(3) A sharp front formed in the solar wind ahead of the magnetosphere, marked with a sudden slowing-down of the flow near the Earth. It is quite similar to the shock forming ahead of the wing of a supersonic airplane. After passing near the Earth, the slow-down flow gains speed again, to the same value as the surrounding solar wind.

bow-shock the magnetic shock generated by the earth as it moves through the interplanetary plasma while dragging along a dipole magnetic field.

breakdown voltage if the gas pressure within a discharge tube is within a range 10^{-4}-10^{-2} m Hg and the applied voltage V between the electrodes is slowly increased, a discharge will occur eventually when a current I, initially transient, will pass through. The V-I characteristics in a gas discharge has been extensively studied. The various regions of discharge are: **Townsend discharge or dark discharge, glow discharge and arc discharge.**

The condition of breakdown of a gas in a discharge tube was first studied by Townsend. Breakdown is said to occur when the discharge through the gas just starts. It is dependent on the value of E/p (where E = V/d, is the electric field and d is the distance between the electrodes of the discharge tube. Townsend criterion for breakdown is given by $\gamma(e^{\alpha d} - 1) = 1$, where α = Townsend's first coefficient, defined as the number of ion-pairs on the average that can be created by an electron in traversing a distance of .01m and γ = Townsend's second coefficient, defined as the number of secondary electrons ejected from the cathode by an ion (produced in the volume by an electron) reaching it and d = distance between the electrodes. In the theory of the Townsend discharge, an important law, known as Paschen's law, is written as $V_b = f(pd)$, where V_b = breakdown voltage, p = gas pressure, d = distance between the electrodes. V_b plotted against pd shows a minimum—that can be explained in terms of ionization probability at high and low pressures.

breakeven the condition in which the fusion power produced by the plasma equals the heating power put into the plasma. For a fusion power reactor to be economically useful, a burning plasma must produce much more fusion power than it consumes, thereby significantly exceeding the breakeven point.

breakeven (commercial, engineering, scientific and extrapolated) (1) commercial breakdown is when sufficient fusion power can be converted into electric power to cover the costs of the fusion power plant at economically competitive rates.

(2) engineering breakeven is when sufficient electrical power can be generated from the fusion power output to supply power for the plasma reactor plus a net surplus without the economic considerations.

(3) scientific breakeven is when the fusion power is equal to input power, i.e. Q = 1.

(4) extrapolated breakeven is when scientific breakeven is projected for actual reactor fuel (e.g. deuterium and tritium) from experimental results using an alternative fuel (e.g. deuterium) only by sealing the reaction rates for the two fuels.

breeder a term sometimes used to indicate that component of a fusion power plant used to "breed" or produce via nuclear reactions, tritium from the energetic neutrons released, for use as fuel in the power plant.

Brehmsstrahlung radiation electromagnetic radiation produced by the rapid deceleration of an electron during a close approach to an atomic nucleus. The radiative loss due to this "breaking effect" increases rapidly with the energy of the electron. The energy loss at each encounter is radiated as a single photon. The Bremsstrahlung forms an important constituent of cosmic radiation.

the important aspect of the fusion device that has to be paid attention to, is to prevent loss of energy due to radiation from fast charged particles of the plasma. Particularly important is the electron-ion Bremsstrahlung radiation. For a fusion device to be self-maintaining, the power produced by the fusion reaction must be greater than the power wasted by radiation.

burn an ignited plasma is said to be "burning".

burst a transient enhancement of the solar radio emission usually associated with an active region or flare.

butterfly diagram a plot of observed solar active region latitudes vs time. This diagram, which resembles a butterfly, shows that the average latitude of active region formulation drifts from high to low latitudes during a sunspot cycle.

C

capacitively coupled discharge plasma plasma created by applying an oscillating radio-frequency potential between two electrodes. Energy is coupled into the plasma by collision between the electrons and the oscillating plasma sheaths. If the oscillating frequency is reduced, the discharge converts into a glow discharge.

carbon burning in the process of evolution of stars, the first stage is hydrogen burning. When He^4 is used up, the next phase is carbon burning. Burning of two carbon atoms produces Mg^{24*} and other atoms.

carbon nitrogen cycle in this process of evolution of a star by hydrogen burning hydrogen burns by two different processes into helium. In the carbon-nitrogen cycle, protons burn in He^4.

causality condition the state of the plasma is not uniquely determined by the incident wave, but further conditions have to be imposed to specify the state. These conditions express the requirement that the plasma is at rest before the incident wave has hit the plasma. the condition that no radiation should emerge before the incident waves have reached the plasma is called "causality condition".

caviton when an electron plasma wave goes nonlinear, the dominant new effect is that the ponder motive force of the plasma waves causes the background plasma to move away, causing a local depression in density called a caviton.

CCD—charged couple device a very sensitive electronic method of detecting light.

CDX-U Current Drive Experiment—Upgrade. a small spherical tokamak at Princeton. CDX-U *Home Page*

centimeter burst a solar radii burst in the centimeter wavelength range (1 to 10 cm).

central electron density the electron density is the number of electrons per unit volume. These change as one move from the tokamak axis to the edge. The central electron density (the density in the centre of the plasma) is important since it is in the centre where most fusion reactions take place.

centre-of-mass velocity let us consider binary elastic collisions that occur in the gas. Let m and m_1 be the masses of the colliding particles and let $\vec{V}$, $\vec{V}_1$ and $\vec{V}'$, $\vec{V}'_1$ be the velocities of m and m_1 before and after the collision. The **centre-of-mass velocity** (before the collision) is defined as

$$\vec{V}_0 = \frac{m\vec{V} + m_1\vec{V}_1}{m + m_1}$$

while the relative velocity $\vec{V}_R$ (before the collision) is $\vec{V}_R = \vec{V}_1 - \vec{V}$. Because of the momentum-and energy conservation laws, we can show that after an elastic collision, the centre of mass velocity is **unchanged**, while the relative velocity **changes in direction but not in magnitude.**

To prove these statements, we first note that, from conservation of momentum, we have $(m + m_1)\vec{V}_0 = m\vec{V} + m_1\vec{V}_1 = m\vec{V}' + m_1\vec{V}_1'$ $= (m + m_1)\vec{V}_0'$, so that $\vec{V}_0'$ is the centre-of-mass velocity **after the collision.** It clearly equals $\vec{V}_0$. Next, if we let μ be the reduced mass, then

$$\mu = \frac{mm_1}{m + m_1}$$

then by direct calculation using these equations, we have

$$^1/_2\{(m + m_1)V_0^2 + \mu V_R^2\} =$$

$$^1/_2\left\{(m + m_1)\left[\frac{m\vec{V} + m_1\vec{V}_1}{m + m_1}\right]^2 + \frac{mm_1}{m + m_1}(\vec{V}_1 - \vec{V})^2\right\}$$

$$= {}^1/_2\left\{\frac{(m\vec{V} + m_1\vec{V}_1)^2}{m + m_1} + \frac{mm_1}{m + m_1}(V_0^2 + V^2 - 2VV_1)\right\}$$

$$= {}^1/_2\left\{\frac{m^2V^2 + m_1^2V_1^2 + 2mm_1VV_1 + mm_1V_1^2}{m + m_1} \frac{+ mm_1V^2 - 2mm_1VV_1}{}\right\}$$

$$= {}^1/_2\,\frac{mV^2(m + m_1) + m_1V_1^2(m + m_1)}{m + m_1}$$

$$= {}^1/_2(mV^2 + m_1V_1^2).$$

Hence, by conservation of kinetic energy, the expression on the right is conserved in a collision so that

$$^1/_2\left\{(m + m_1)\left[\frac{m\vec{V} + m_1\vec{V}_1}{m + m_1}\right] + \frac{mm_1}{m + m_1}(\vec{V}_1 + \vec{V})^2\right.$$

$$= {}^1/_2\left\{(m + m_1)V_0^{12} + \frac{mm_1}{m + m_1}(V_1{}^1 + V')^2\right\}$$

Since we have already shown that $\vec{V}_0' = \vec{V}_0$, this reduces to $(\vec{V}_1 - \vec{V})^2 = (\vec{V}_0' = \vec{V}')^2$, so that, $|\vec{V}_1' - \vec{V}'|$, the magnitude of the relative velocity after the collision, exactly equals $|\vec{V}_1 - \vec{V}|$ collision plane—the plane formed by $\vec{V}_1 - \vec{V}$ and $\vec{V}_1' - \vec{V}'$.

cesium plasma much work has been done using gases "seeded" with elements like cesium that have a low ionization potential, so that the total value of n_e (and σ) is somewhat higher.

the thermionic converter uses a cesium plasma between two electrodes, one heated to emit electrons and the other cooled. Because of the cesium plasma very high currents are produced and a significant portion of the thermal energy applied to the cathode is extracted as electric energy.

chain reaction rapidly accelerating multiple fission reactions resulting from the production of several neutrons by each fission reaction. An uncontrolled chain reaction within a sufficiently large amount of fission fuel (critical mass) can lead to an explosive energy release.

Chandrasekhar he has investigated the stability of hydromagnetic coutle flows (often used to analyze the effect of hydromagnetic forces on boundary layers) for an applied field in the y-direction, **not** in the z-direction. the theoretical predictions have been confirmed experimentally.

Chapman-Enskog method it is a method for truncating the moment equation in which the distribution function is expanded.

charge density charge per unit volume, $\rho = e\,(n_i - n_e)$

charge exchange recombination spectroscopy neutral atoms in the plasma (from, for example, a neutral beam) donate electrons to fully ionized impurity ions, producing hydrogen-like ions. As the electrons decay from excited states they emit photons from which the impurity temperature, rotation and density can be measured using conventional spectroscopy.

charge inequality $n_e - n_i = n_1$, say, is known as charge inequality.

charge transfer it can be represented as any of the following symbolic equations:

$A^+ + B \rightarrow A + B^+$, $A^{++} + B \rightarrow A^+ + B^+$, $A^{++} + B \rightarrow A + B^{++}$

Child-Langmuir law description of electron current flow in a vacuum tube when plasma conditions exit that result in the electron current scaling with the cathode-anode potential to the 3/2 power.

chromosphere (1) a reddish layer in the sun's atmosphere, the transition between the photosphere and the corona.

(2) the layer of the solar atmosphere above the photosphere and beneath the transition region and the corona. the chromosphere is the source of the strongest lines in the solar spectrum and is the source of the red colour seen around the rim of the moon at total solar eclipses.

classical confinement in this, particle and energy transport occur via classical diffusion, best possible case for magnetically confined plasmas.

classical diffusion in plasma physics, diffusion due solely to scattering of charged particles by coulomb collisions stemming from electric fields of the particles. In classical transport (i.e. diffusion) the characteristic step size is one gyro radius (Larmor orbit) and the characteristic time is one collision time.

CLEO—Closed Line Experiment a culham facility to investigate stellarator and TOKAMAK plasmas at large aspect ratio (about 7). Now dismantled.

CMA (Clemmow-Mullaly-Allis) diagram in this diagram, cold plasma waves have been clearly divided into different classes. To often cut-offs and resonances for different high and low frequency waves and for different values of angles θ between $\vec{K}$ and $\vec{B}$, help of the CMA diagram is taken.

coefficient of diffusion when there is a gradient of the number density $\vec{\nabla} n$ between two adjacent layers of a plasma, the number of excess particles per unit area per unit time transported from higher density to lower is known as the **particle flux** Γ, given by $\vec{\Gamma} = -D\vec{\nabla}n$, D being called the coefficient of diffusion. In the M.K.S. system D is expressed as m^2/sec. The coefficient of diffusion for electrons D_e is much greater than that of ions D_i.

coefficient of diffusion in the velocity space the term $<\partial\vec{V}\partial\vec{V}>$. in the Fokker-Planck equation (see Fokker-Planck equation) is called coefficient of diffusion in the velocity space.

coefficient of dynamical friction the term $<\partial\vec{V}>$ in the Fokker-Planck equation is called coefficient of dynamical friction.

coefficient of reflection assuming an isotropic velocity distribution of particles within the magnetic mirror, the coefficient of reflection, R, is defined as the fraction of particles, approaching the mirror per second, that are reflected.

foil single or multiple-turn winding of a conductor such as copper wire. When the conducting windings carry current, a magnetic field is produced which can be used for shaping or stabilizing a plasma, or for inducing plasma current.

cold plasma in a "cold" plasma (temperature $\approx$ a few eV, compared to a "hot" plasma having temperature $\approx$ a few hundred eV) with low electrical conductivity, collision results in leaking of the magnetic field through the plasma, or in other words, in the diffusion of plasma across the magnetic field, so that the freezing concept breaks down.

collision frequency the concept of a collision frequency is probably the most important one in plasma physics (and the physics of gases in general) when it comes to assessing the significance of the individual physical processes. It is defined as

$\nu_c = N\sigma_c(V)\, V,$

where N = volume density (i.e. number per unit volume) of the background medium and $\sigma_c(V)$ the cross-section of the particle with velocity V for the type of collision being considered (e.g. Coulomb collisions, radiative recombination, collision excitation).

Despite the random nature of collisions, ν_c can be considered as an exact quantity because the large number of particles usually assures that the average is very sharply defined within relatively short time scales and small volumes.

collisionality a measure of how frequently collisions occur in a TOKAMAK plasma.

A collisionality of unity corresponds to a trapped particle performing a single banana orbit before being scattered.

collisionless plasma model model of plasma where the density is low enough and/or the temperature is high enough, that collision can be neglected because the plasma time scales of interest are shorter than the particle collision time.

collision when a particle of a gas or a plasma happens to come near another particle (or several of them), the magnitude and/or direction of their velocities and, sometimes, even their internal energies may change. it is then said that a collision has taken place. Collisions play the main role in the kinetic theory of gases. In a plasma, collision is secondary and is considered to be only a small perturbation. Collision between molecules of a gas are like collision between two billiard balls and, therefore, mostly binary. In contrast, collisions between plasma particles are of an entirely different character. Here a single electron can simultaneously interact with many of its neighbours, near and far, through long range coulomb field.

collision, hard the force between two neutral molecules is of short range and the collisions that occur due to their thermal motions are random, sudden and strong. This type of collision is "hard" and the angle of scattering may be quite large.

collisions, inelastic inelastic collision is one in which both the internal and kinetic energies of the interacting particles, e.g., electrons, ions, atoms, molecules, ionized molecules or atoms, excited molecules or atoms, photons etc. can participate in this type of collision.

Inelastic collisions may be of many kinds, e.g. excitation, ionization, recombination, charge transfer, attachment, dissociation, de-excitation etc.

collisions, "soft" for two charged particles, compared to the neutral particles, the force is long range and each charged particle can continuously and simultaneously interact with its innumerable neighbours. As a result, the path of a charged particle cannot be divided into straight lines, as obtained for a gas molecule by joining two consecutive points where collisions occurred. Hence, in this case, collisions are not sudden or strong and, therefore, "soft".

collisions, elastic a collision in which the total kinetic energy of the interacting particles remains unaltered is known as the elastic collision. Therefore, in this type of collision, the internal energies of the particles also remain the same. Such collision may occur between two electrons, two ions, an ion and an electron, and electron and a molecule or between two neutral molecules.

comet a small body of ice and dust which orbits the Sun.

compass (-C)(-D) compact assembly; the Culham conventional aspect ratio TOKAMAK facility.

It has a similar magnetic geometry to JET and, therefore, plays a crucial role in scaling experimental results through JET to ITER.

compatibility for arbitrary initial values of the field quantities, which obey the condition equations, it should be possible to find solutions for all subsequent which again obey the condition equations.

completeness the solutions must be uniquely determined by the initial values.

conductivity the heat flux $\vec{Q}$, representing excess thermal energy per unit area per unit time, is given by

$$\vec{Q} = -K\vec{D}T$$

where K is conductivity. In the M > K > S > system it is expressed as watt/m/^{0}K. Current density $\vec{j} = \sigma\vec{E}$, where σ = electrical conductivity. Σ is expressed in terms of mobility μ by $\sigma = ne\mu$.

conductivity d.c. the equation $\vec{J} = \sigma(\vec{E} + \vec{V} \times \vec{B})$ with $\sigma = me^2/m\nu$ is often referred to as a **generalized Ohm's law** and σ is called **the d.c. conductivity of the plasma.** In addition, when $\nu \to 0$, so that $\sigma = \infty$, the current $\vec{J}$ can remain finite only if

$\vec{E} + \vec{V} \times \vec{B} = 0$. Hence the equation $\vec{E} + \vec{V} \times \vec{B} = 0$ often called Ohm's law for a plasma with infinite conductivity.

conductivity tensor the current produced by a stationary electric field in the presence of a d.c. magnetic field (in the Z-direction) is given by

$J_x = \sigma_\perp E_x + \sigma_H E_y$, $J_y = -\sigma_H E_X + \sigma_\perp E_y$, $J_Z = \sigma_0 E_Z$.

Hence this may be written as

$\vec{J} = \vec{\sigma}.\vec{E}$ with

$$\sigma = \begin{bmatrix} \sigma_\perp & \sigma_H & 0 \\ -\sigma_H & \sigma_\perp & 0 \\ 0 & 0 & \sigma_0 \end{bmatrix}$$

the d.c. conductivity tensor.

conductivity hall this conductivity governs the current flow perpendicular to both $\vec{E}$ and $\vec{B}$ and is given by $\sigma_H = \sigma_0 \omega_c \nu/(\omega_c^2 + \nu^2)$, where σ_0 = d.c. conductivity of the plasma. $\sigma\perp$ is called the perpendicular conductivity, given by $\sigma\perp = \sigma_0 \nu^2/\omega_c{}^2 + \nu^2$, which governs current flow in the direction of $\vec{E}$ when $\vec{E}$ is perpendicular to $\vec{B}$.

confinement one of the important branches of plasma research is in the area of controlled thermonuclear fusion. Essentially, the goal of this programme is to make a miniature star on earth, generating energy by means of the same reaction that takes place in the sun. The fundamental process is that of the nuclear fusion of heavy isotopes of hydrogen into helium, with the subsequent release of energy. In order to achieve the fusion process, the reactants must be heated (to some millions of degrees) and then kept together in sufficient concentration for a period of time to allow the reaction to become exothermic.

Since the plasma must be kept both hot and in a concentrated form, the first problem is the choice of a suitable means of production and storage (confinement) of the plasma. The high energies can more easily be achieved under low pressure (between 10^{-2} and 10^{-3} Torr), so a vacuum system is usually required. Then, the extremely high energies of the particles must not be allowed to dissipate by, for example, allowing the plasma to come in contact with the walls of the vacuum chamber, so some form of confining scheme to keep the plasma away from the walls must be used. It should be noted that nature has achieved. This result already in the Sun and stars. For this "controlled fusion reactor" the vacuum system is outer space itself and the confinement scheme is the self-gravity of the stars. on earth, the mass of fusion reactors will obviously be far to small to make use of this method.

The approach for confinement up to now has been to use the electrical properties of plasma particles. Once ionized, the plasma particles are affected by electric and magnetic fields. Should a scheme be made which then could confine a thermonuclear plasma so that it does not interact with the walls and could keep it concentrated for a long enough time, the problem will have been essentially solved. This method has used externally applied electric, magnetic, and/or electromagnetic fields to attempt to produce adequate confinement.

Controlled fusion can be achieved by having a large number of hot plasma particles together for a short period of time or a smaller number of particles together for a longer period of time. That is, the product of plasma density times confinement time is roughly a constant. As a consequence, many fusion experiments that operate in the high density short **confinement time** regime can utilize a pulsed mode of operation, which often allows the apparatus to be less massive.

If n is the number density in m^{-3} and τ be the **confinement time** in sec., then the product $n\tau \geq 1.5 \times 10^{22}$ $m^{-3}s$ for the high probability of fusion. This condition was first given by J. D. Lawson. The condition is called the **Lawson criterion.**

confinement schemes many confinement schemes have been suggested. They fall into one of the following categories:

(a) pinch devices, in which a current produces a force tending to compress the cross-sectional area of the plasma carrying it, so that the plasma is both heated and confined.

(b) mirror machines, which are linear devices with an axial magnetic field to keep particles from the wall and also to reduce the number of charged particles that escape at each end.

(c) sellarator-type machines, where again the confinement of the plasma is by an externally applied, largely axial field but with a tube closed on itself (typically like figure of eight) to eliminate losses of particles at the ends.

The major problem in the fusion programme is that all these confinement schemes are, in one way or another, unstable and small fluctuations from the desired configuration can quickly lead to a rapid escape of the particles from their "magnetic bottle". Interestingly, since thermonuclear plasmas are highly rarefied, their total heat content is fairly low, hence when the hot plasma escapes and comes in contact with the surrounding vessel, the walls of the vessel are not damped, but the plasma itself is quickly cooled to temperatures below the threshold for fusion.

confinement time (see confinement)

conservation of charge in any interaction charge is conserved. The equation of continuity, namely $\vec{D}.\vec{J} + \partial\rho/\partial t = 0$ expresses conservation of charge (ρ = charge density, $\vec{J}$ = current density) of energy. The sum of kinetic and potential energies is a constant.

constitutive equations for many media $\vec{D}$ and $\vec{B}$ are related to $\vec{E}$ and $\vec{H}$ by the simple constitutive equations $\vec{D} = \in \vec{E}$ and $\vec{B} = \mu\vec{H}$, where $\in$ = dielectric constant (farads/m) and μ = magnetic permeability (henrys/m).

continuity equation the equation

$$\frac{\partial n(\vec{r},t)}{\partial t} + \vec{\partial}.\{n(\vec{r},t)\vec{u}(\vec{r},t)\} = 0$$

is known as the first moment equation and is called the equation of continuity in hydrodynamics. It states that the number of particles is conserved; being multiplied by e or by m, it expresses the conservation of charge or mass, respectively.

continuum optical radiation arising from broadband emission from the photosphere.

continuum model it is a model for the plasma, the so-called hydrodynamic or continuum model, in which the plasma is treated as an interacting mixture of electron ion, and neutral molecule fluids (or gases), each with its separate set of mass, momentum and energy-transport equations.

controlled thermonuclear fusion laboratory experimental plasmas in which light nuclei are heated to high temperatures (millions of degrees) in a confined region that result in significant enough fission reactions under controlled conditions to be able to produce energy.

controlled thermonuclear reactions fusion process is possible only when the two interacting particles (D, D or D, T) can come near each other within a distance of the order of 10^{-14}m. This means that they will have to overcome the mutual repulsive coulomb barrier acting between them due to positive charges on both and come within the distance of the above order. And this is why their relative velocity should be very large. That is, the temperature of the medium must be very high. For example, mutual potential energy between two deuterons is e^2/R and the average K.E. is 3/2KT. For the fusion process, $3/2KT = e^2/R$. Or $T = 2/3\ e^2/KR \approx 10^9\ {}^0K$, when $R \approx 10^{-14}$m.

Therefore, the necessary condition for the occurrence of fusion process at a high rate is that the medium must be very hot. This is why the term thermonuclear reaction is used.

convection the bulk transport of plasma (or gas) from one place to another in response to mechanical forces (for example, viscous interaction with the solar wind) or electromagnetic forces.

convective instabilities plasma wave's amplitude increasing as the wave propagates through space without necessarily growing at a fixed point in space.

converging magnetic field flux ϕ over the circular orbit of a particle moving perpendicular to the field is given by $\phi = \pi r^2 B$. In a converging magnetic field (B increasing) the particle orbit is compressed (r decreasing) in such a way as to maintain ϕ constant. Using the constancy of the magnetic moment μ and the principle of conservation of energy, it can be shown that the converging magnetic field can reflect particles and can behave as a **magnetic mirror.**

corona (1) the outermost layer of the solar atmosphere, characterized by low densities. it is not visible from the earth except during a total eclipse of the sun or by use of a special telescope called a coronagraph.

(2) the outermost layer of the sun's atmosphere, visible to the eye during a total solar eclipse; it can also be observed through a coronagraph spectral filters and best of all by X-ray cameras aboard satellites. The corona is very hot, about 2 million degrees Centigrade and is the source of the solar wind.

(3) The very hot outer layer of the Sun's atmosphere, composed of highly diffused, superheated, ionized gases and extending into interplanetary space. The hot gases in the solar corona form the solar wind.

(4) The outermost part of a star's atmosphere, the corona characterized by high temperatures and low densities relative to the stellar photosphere. The corona is home to many plasma phenomena, the origin of the high temperature

corona it is a natural plasma having electron density $\approx 10^{14} m^{-3}$, electron temperature $\approx 10^6\ ^0K$ and Debye length $\approx 0.5 \times 10^{-2} m$.

coronagraph an optical device that makes it possible to observe the corona at times other than during an eclipse. A simple lens focuses an image of the sun into an occulting disc that prevents the light from the solar disc from proceeding further along the optical path, effectively providing an artificial eclipse.

coronal hole (1) an extended region of the corona, exceptionally low in density and associated with unipolar photospheric regions having "open" magnetic field topology. Coronal holes are largest and most stable at or near the solar poles and are a source of high speed solar wind. Coronal holes are visible in several wavelengths, including solar X-rays and the He 1083mm emission line.

coronal loops a typical structure of enhanced corona observed in EUV lines and soft X-rays. They are sometimes related to H alpha loops. Coronal loops represent "closed" magnetic topology.

coronal mass ejection (CME) (1) A vast region of hot, dense and high speed solar wind propagating away from the sun.

(2) A huge cloud of hot plasma, expelled sometimes from the sun. it may accelerate ions and electrons and may travel through interplanetary space as far as the Earth's orbit and beyond it, often preceded by a shock front. When the shock reaches the Earth, a magnetic storm may result.

(3) A transient outflow of plasma from or through the solar corona. CMEs are often, but not always, associated with erupting prominences, disappearing solar filaments and flares.

(4) Aporadic ejection of plasma and magnetic field from the Sun, usually observed by optical telescopes with the Sun blocked off. An event can eject tens of trillions of tons of predominantly hydrogen plasma at hundreds of kilometers per second. Also called solar mass ejection.

coronal neutral line the line in the corona that separates solar magnetic fields of opposite polarity. It is calculated from solar observations of the photospheric magnetic field. Extension of the neutral line radially outward by the solar wind defines the current sheet in the heliosphere.

coronal rain material condensing in the corona and appearing to rain down into the chromosphere as observed in H alpha at the solar limb above strong sunspots.

coronal streamer (1) A feature of the white light corona that looks like a ray extending away from the Sun out to about one solar radius, having an arch-like base.

(2) A large scale structure in the white light corona often overlying a principal inversion line in the solar photospheric magnetic fields.

cosmic ray (1) electromagnetic radiation of extremely high frequency and energy, cosmic rays usually interact with the atoms of the atmosphere before reaching the surface of the Earth. Some cosmic rays come from outside the solar system while others are emitted from the Sun and pass through holes in the corona.

(2) Nuclear and subatomic particles moving through space at speeds close to the speed of light. They are thought to come from stars in the Milky Way galaxy.

coulomb barrier (See Controlled Thermonuclear Reactions).

coulomb collision particle collisions where the coulomb force is the governing force that result in the deflections of the particles away from their initial paths.

coulomb crystal regular crystal-like arrangement of electrically charged particles in potential traps, in plasmas, or in colloidal suspensions.

coulomb explosion when a sufficiently intense laser irradiates a group of atoms (gas cluster, object target, etc.), the electric field of the laser beam can drive some or all of the electrons off the atoms. With the electrons gone, the resulting group of ions explodes due to the coulomb repulsion of the positive charges.

coulomb gas the role of the coulomb force is dominant in plasma. Therefore, a plasma is also called a coulomb gas.

crab nebula various plasma theories have been used to explain the acceleration of cosmic rays. Although the stars in a galaxy are not charged, they behave like particles in a plasma and plasma kinetic theory has been used to predict the development of galaxies. Radio astronomy has uncovered numerous sources of radiation that most likely originate from plasmas. The Crab Nebula is a rich source of plasma phenomena because it is known to contain a magnetic field. It also contains a visual pulsar. Current theories of pulsars picture them as rapidly rotating neutron stars with plasma emitting synchrotron radiation from the surface.

criteria of plasma there are several criteria for calling an ionized gas a plasma :

(i) a fluid containing sufficiently numerous free charged particles causing its dynamical character to be dominated by electromagnetic fields.

(ii) a quasi-neutral gas of charged and neutral particles.

(iii) a fully ionized gas of low density.

(iv) a coulomb gas.

(v) a fourth state of matter resembling gaseous state at high temperature.

(vi) a system of charged and neutral particles satisfying certain criteria, e.g. $L >> \lambda_D >> n_e^{-1/3}$ and $\nu < \omega_p$.

critical frequency in the ionosphere, the electron plasma frequency increases with altitude until a maximum value ω_{cr}—the so-called critical frequency—is attained. This maximum value is reached, generally, at a height of about 300 km and, thereafter, the plasma frequency falls off slowly with increasing altitude. Hence a radio wave transmitted vertically upward with frequency ω below ω_{cr} is reflected at an altitude when $\omega_e = \omega_i$.

cryogenics engineering processes carried at low temperatures (usually below 120K). Cryogenics in fusion is used for cooling the superconducting magnets (if present) and for cryopumping.

current density it is given by $\vec{j} = n_i e\vec{V}_i - n_e e\vec{V}_e = ne(\vec{V}_i - \vec{V}_e)$, since $n_e \approx n_i = n$.

Or, $\vec{j} = ne(\mu_i - \mu_e)\vec{E}$, since $\vec{V}_e = -\mu_e\vec{E}$ and $\vec{V}_e = -\mu_e\vec{E}$.

Or, $\vec{j} = ne\mu_e\vec{E}$ (since $\mu_e >> \mu_i$)

$= \sigma\vec{E}$, where $\sigma = ne\mu_e$, the electrical conductivity of plasma.

current distribution the variation of plasma current density within the plasma, usually expressed as a function of the distance from the magnetic axis (see profile).

current drive (non-inductive) a method of driving plasma current (in a TOKAMAK) that does not depend on transformer action (e.g. by using RF waves or neutral beams); necessary for a continuously operated power plant, since transformer action is cyclic. Also being applied to control instabilities and to optimise confinement. See electron cyclotron current drive, fast wave current drive, helicity injection, ion cyclotron current drive, lower hybrid current drive.

current ramp-up (down) the increase (decrease) of plasma current—either at the start of operation (ramp-up) or during operation—to modify the current profile for performance investigations.

curvature drift if the lines of force of an inhomogeneous field $\vec{B}$ are curved having the radius of curvature R, and if the charged particle moves along the curved lines of force with a velocity $\vec{V}_{11}$, then a drift known as the curvature drift occurs, which is given by $|\vec{V}_R| = V^2_{11}/R\omega_c = mV^2_{11}/ReB = 2\omega_{11}/eR^2B^2\,|\vec{R}\times\vec{B}|$. The direction of $\vec{V}_R$ is along $\vec{R}\times\vec{B}$, that is, along Y (figure)

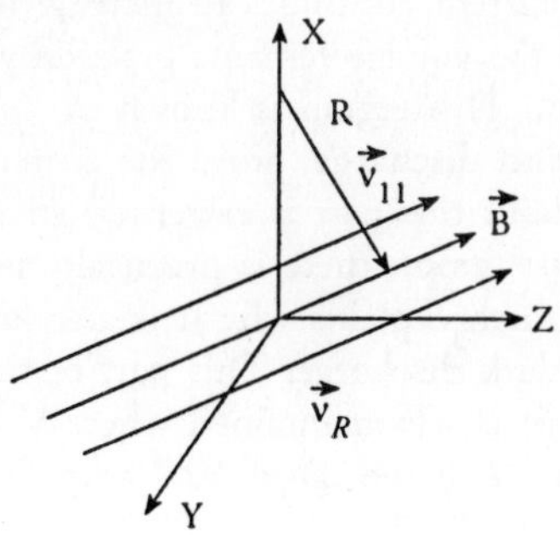

cusped magnetic fields difficulties experienced with the mirror machines with, regard to loss due to diffusion in the perpendicular direction as well as the ends, are partly removed by reversing the current in one of the coils. The cusped field will look like that shown in the figure.

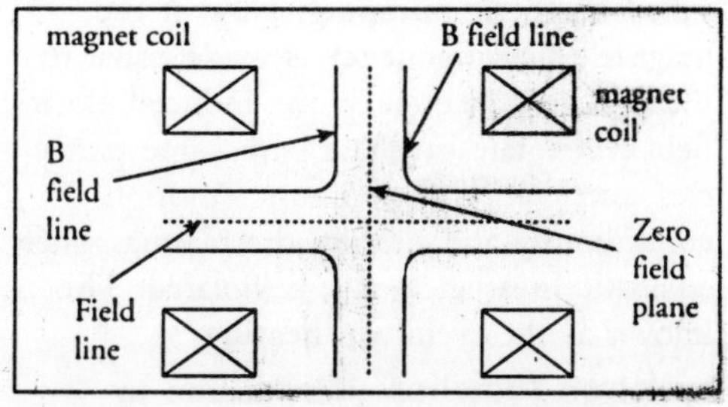

cut-off when $\omega < \omega_p$, the group velocity Vg is imaginary, which means that the wave cannot propagate through the plasma. This is known as the cut-off. In the case of cut-off, waves are generally reflected; that is why cut-off point is alternatively known as the reflection point.

cyclotron angular frequency this is also known as Larmorf frequency or gyro-frequency ω_c given by $\omega_c = V\perp/r = eB/m$.

cyclotron damping this is a strong form of damping which occurs when some particles experience a perturbing force, which oscillates at their cyclotron frequency. Both electron cyclotron damping and ion-cyclotron damping can occur.

In this type of damping, energy is lost, that is, the energy of the electromagnetic field of the wave is converted into the random thermal energy of the particles. This process is used to heat the plasma. In cyclotron damping, the wave energy can be converted into the energy of the particles in a relatively short distance. since the transfer of energy from electron gas to ion gas is very slow, the ion cyclotron damping is used in most cases, so that the loss of energy by the waves is a straight-forward gain of energy by the ions.

cyclotron heating there are situations where the plasma in the trap can be heated when adiabatic invariance of μ is violated. Thus, if the magnetic field in the mirror trap is periodic, the transverse velocity $\vec{V}\perp$ would oscillate, though without any gain of energy due to this. However, if collisions occur, adiabatic invariance of μ will not hold and plasma will be heated. Specially, an oscillating particle during its compression phase can transfer a fraction of its gyration energy into longitudinal V_{11} energy. This process is called **magnetic pumping**. Now, if the magnetic field frequency is made equal to the cyclotron frequency, the induced electric field will rotate in phase with some particles, accelerating their Larmor motion continuously and heating the plasma, when adiabatic invariance of μ is violated. This is known as the **cyclotron heating**.

cyclotron radiation the radiation by orbiting particles in a magnetic field.

cyclotron radius if a charged particle moves at right angles to $\vec{B}$, the orbit of the particle is a circle. The centre of this circle is known as the **guiding centre** and the radius r the **radius of gyration** or **gyro-radius** or **Larmor radius** (in honor of the British physicist Larmor) or **cyclotron radius**.

cylindrical approximation an approximation to the true TOKAMAK geometry in which the torus is cut and straightened, so that the toroidal direction becomes the cylinder axis. There are two directions of symmetry: Along the axis (the "toroidal" direction) and About the axis (the "poloidal" direction).

cryopump a device (containing a coolant such as LHe or LN2) for trapping condensable gases from neutral beams or the plasma. When a layer of frost has built up the pump is regenerated. it is a form of vacuum pump. See argon frosting, pumping speed, cryosorption.

D

d.c. conductivity it is given by $\sigma = ne^2/m\nu$ where symbols have usual meanings. In the absence of collision, $\nu \to 0$, $\nu \to \infty$, —d.c. conductivity of plasma is infinity in the absence of collisions.

DP machine ion acoustic shock waves have been generated by Taylor, Baker and Hazi. To do this, a new plasma source, the DP (or double-plasma) device, was invented.

damping of plasma oscillations in deriving the dispersion relation for the plasma oscillations, namely, $\omega^2 = \omega_p^2 + c^2k^2$, one usually neglects the interaction between the electrons and the neutral particles. This interaction results in the fading of the plasma oscillations, because the energy of the electron oscillations is transferred to the neutral particles.

damping, collisionless in connection with Landau damping a negative imaginary part appearing in ω leads to temporal decay. this decay term which arises in the absence of collisions is called the **Landau collisionless damping factor**.

damping, electron cyclotron and ion cyclotron when the refractive index does not become infinite at the electron and ion resonances, but takes a large complex value, it corresponds to a damping of the same kind as the Landau damping; these effects are called the electron cyclotron damping and the ion cyclotron damping.

dark discharge if one studies the V-I characteristics of discharge in a gas, one obtains several distinct regions where either the voltage V or the current remains constant. For example, in the first part, current is proportional to voltage, that is, Ohm's law is obeyed. In the second part, current remains essentially constant, although the voltage continues to increase. This is known as the saturated current region. In these two parts, the discharge is not self-maintained. In the third part, with the further increase in voltage, current increases, but then a region is found in which current continues to increase rapidly even if the voltage remains practically constant. This region is known as **Townsend discharge**. Since the current throughout this part is extremely small (less than 10^{-9} amp), there is practically no glow, which explains why it is also known as the **dark discharge**. This part of the discharge is self-maintained where $V \sim 10^3$ volts and I in the range 10^{-3} amp to 10^{-6} amp.

Debye length also known as the Debye radius or Debye Screening distance and given by $\lambda_D = (\varepsilon_0 kT/n_0 e^2)^{1/2}$ (usual meaning of symbols).

Debye potential the Debye potential ϕ of a charge q is a statistical concept which arises due to the average behavior of the particles surrounding the charge. The potential of the charge q falls off exponentially with distance and is reduced to 1/e times the value at r = λ_D. It is given by

$$\phi = \frac{q}{4\pi\varepsilon_0 r} e^{-r/\lambda_D}.$$ in free space, $n_0 \to 0, \lambda_D \to \infty$

and the potential due to a charge q falls off as 1/e, thus extending to ∞.

Debye screening the Coulomb force among the plasma particles varies as r^{-2}. But that does not mean that in the case of long-range collisions, this force is extended to infinity. Because of the dielectric properties or polarization of the plasma, the ionized particles surrounding a charge offer an electric shield or screen, so to say, and hence the effective electric field of the charge does not extend to infinity, but to a characteristic distance λ_D, known as the Debye length. This length plays a fundamental role in the theory of plasmas.

Debye sphere the spherical region in which the density is increased and in which the potential differs appreciably from zero is called the Debye sphere. Rigorously speaking, it extends to infinity, but it may be visualized as a sphere whose radius is of the order of the Debye length.

Debye-Hüchel radius it is the same as Debye length or Debye radius λ_D.

de-excitation this process is reverse of the process of excitation and can be represented by $A^* \to A + h\nu$. It is spontaneous.

delta-effect an effect which consists of secondary electron emission from the cathode by photons, produced by recombination of electrons and ions in the gas and falling on the cathode.

DEMO a demonstration fusion power plant which will resemble a commercial fusion power station as closely as possible. This stage in the progress from present-day machines to commercial fusion power is expected to be constructed after a Next Step device.

density in plasma physics "density" nearly always refers to the number density—the number of particles per unit volume (as opposed to the mass density—mass per unit volume).

density limit plasma disruptions occur within TOKAMAKs if the electron density is too high. See Hugill diagram, Greenwald density.

detailed balance in equilibrium, if plasma could be isolated from its surroundings, the rate of emitting transitions could be exactly balanced by the rate of absorbing transitions, so that energy is conserved. That is, the number of emitting transitions where electron momentum changes from $\vec{p}'$ to $\vec{p}$ must be exactly balanced by the numbers of absorbing transitions, which change the momentum from $\vec{p}$ to $\vec{p}'$. This statement is the principle of detailed balance.

Under these conditions, it is possible to obtain a relationship between the emission and absorption coefficient if collective efforts of the plasma can be neglected. In order to do this, however, we must first consider the photon distribution. a plasma will have, at any instant, not only a distribution of particle momenta and positions, but also a distribution of energies of radiation photons.

deuterium a stable isotope of hydrogen, whose nucleus contains one proton and one neutron. Deuterium plasmas are used routinely in present-day experiments.

deuterium-helium-3 reaction an alternative fusion reaction to deuterium-tritium :

$D + {}^3He \to p + \alpha + 18.4$ MeV

Since the fuel is not radioactive and few neutrons are produced (from subsequent reactions between the daughter products), this reaction has safety and environmental advantages over D-T. However, it is harder to achieve, with higher temperatures (three times), densities and confinement times required for power plant operation.

deuterium-tritium reaction the most likely fusion reaction to be used in future fusion power plants :

$D + T \to n + \alpha + 17.6$ MeV

Deuterium-tritium plasmas have already been used to produce significant amounts of fusion energy in both the JET and TFTR devices.

diagnostics engineers or experimenters working with ordinary gases have no difficulty in determining the physical properties and state of the gas. Any thermometer or pyrometer can measure the temperature of the gas; a manometer the pressure; a flow meter, the stream velocity; finally, highly developed chemical and physicochemical methods of gas analysis yield a determination of the chemical composition. Plasma represents a different situation. Every measurement is a problem. There are many cases in which temperature measurements by different methods on the same plasma yield results which disagree by a factor of ten. Also, we find that many experimenters working with plasmas do not even know such basic quantities as the charged particle density with precision.

The reason for this state of affairs is that the determination of the physical characteristics of plasmas, in contrast with gases, does not reduce to a simple measurement. Methods for determining the temperature, density and composition of plasmas are the subject of an important part of experimental plasmas physics which has come to be called **diagnostics.** To arrive at an opinion concerning the condition of a plasma by reading of instruments is as difficult as trying to establish a diagnosis of an illness by an inspection of a patient. Whereas an ordinary gas responds to the questions of the "doctor" willingly, a plasma only groans like a dumb animal so that the plasma researcher should, perhaps, be compared to a veterinarian.

one of the diagnostic methods is the use of a Langmuir (electrostatic) probe. In the absence of a magnetic field, the electron density and temperature of a plasma can be determined simultaneously by this probe.

diamagnetic current in a magneto plasma, the thermal motion of particles at right angles to the magnetic field is a cyclotron gyration. each of the rotating charged particles forms a circular current in the plane perpendicular to the magnetic field. These current loops possess magnetic moments, which is why they are called magnetization currents. The ions and electrons rotate in opposite directions, but the current they carry are additive. The direction of the current is commonly taken to be the direction of motion of the positive charges. The magnetization current is excited by the external magnetic field. However, the field set up by these currents is opposite in direction to the external field, thereby weakening the latter. In the absence of other currents the plasma behaves as a diamagnetic medium. For this reason, the magnetization current is also known as the **diamagnetic current.**

diamagnetic drift it is given by

$$\vec{V}_D = -\frac{\vec{\nabla} p \times \vec{B}}{enB^2}$$

diamagnetic loop a diamagnetic loop is a device used to measure plasma pressure by detecting the diamagnetic effect. As the plasma is created, the diamagnetic current increases, B decreases inside the plasma and the flux ϕ enclosed by the loop decreases, inducing a voltage, which is then time-integrated by an RC circuit.

dielectric constant ϵ_R, the dielectric constant of the plasma relative to the vacuum value, is given by

$\epsilon_R = \frac{\epsilon}{\epsilon_0} = 1 + \frac{\mu_0 \sigma c^2}{B^2}$, where the symbols have their usual meanings.

dielectric permeability of plasma it is defined by $K_\perp = \frac{m_+ + m_-}{\epsilon_0 B^2} n + 1$

where charge neutrality in plasmas over regions larger than the Debye length is assumed (*i.e.* $\frac{|n_+ - n_-|}{n} << 1$).

dielectric tensor it is defined as

$\vec{K} = 1 + \frac{\Sigma q n_0 \vec{V}}{i\omega\epsilon_0}$, where $\vec{V}$ = mobility tensor.

The dielectric tensor $\in_{il}$ is defined by

$\in_{il} = \partial_{il} + \frac{4\pi i}{\omega}\sigma_{il}$, where the symbols have their usual meanings.

diffusion in a magnetic field the role of collisions in the diffusion process is of one kind in the absence of the magnetic field but of an entirely opposite kind when a magnetic field is present. Thus, in the former case, collision opposes diffusion, while, in the latter, collision is the only process that helps diffusion. When, $\vec{B} \rightarrow 0$ the tendency of the density to become uniformly distributed or equal, in a non-uniform plasma, is restricted by collision among particles ($\therefore$ D = KT/mν, increase in ν decrease D, i.e. collisions impede diffusion). On the other hand, if a strong magnetic field $\vec{B}$ is applied on a non-uniform plasma, particles can only move unopposed parallel to $\vec{B}$ and are resisted strongly in the perpendicular direction, i.e. diffusion is decreased. Diffusion perpendicular to $\vec{B}$ can only occur by collisions, collisions are needed for diffusion. In fact, the relation between transverse diffusion coefficient D⊥ and natural diffusion coefficient D is

$$D\perp = \frac{D\nu^2}{\nu^2 + \omega^2} = \frac{D}{1 + \omega^2\tau^2}$$

where ν = collision frequency = $1/\tau$, τ being the mean free time and ω = eB/m, the Lormor frequency.

diffusion time the time-dependent diffusion equation in one dimension is given

by $D\frac{\partial^2 n}{\partial x^2} = \frac{\partial n}{\partial t}$

It can be utilized in the solution of boundary value problems. In many cases the standard method of separation of variables may be utilized for this purpose.

In three dimensions, the above equation becomes $D\nabla^2 rn = \frac{\partial n}{\partial t}$.

The separation constants in the solution of the above equations will be characteristic lengths or times, often called **diffusion lengths** and **diffusion times**, respectively.

diffusion, Bohm although the theory of diffusion via coulomb collisions had been known for a long time, laboratory verification of the $1/B^2$ dependence of D⊥ in a fully ionized plasma eluded all experimenters until the 1960s. In almost all previous experiments, D⊥ scaled as B^{-1}, rather than B^{-2} and the decay of plasma was found to be exponential rather than reciprocal, with time. Furthermore, the absolute value of D⊥ was far larger than that given by the equation $D = \frac{n_+ n\Sigma kT}{B^2}$. This anomalously poor magnetic confinement was first noted in 1946 by Bohm, Burhop and Massey, who were developing a magnetic arc for use in uranium isotope separation. Bohm gave the semi empirical formula

$$D_\perp = \frac{1KT_e}{16eB} \equiv \tau_B$$

This formula was obeyed in a surprising number of different experiments. Diffusion following this law is called Bohm diffusion. Since D_B is independent of density, the decay is exponential with time. With the flux Γ given by Fick's law and Bohm's formula, one obtains

$$\tau = \frac{R^2}{2Dn} \equiv \tau_B$$

The quantity τ_B is often called the **Bohm time.**

diffusion, forced the drift velocity $\vec{V}$ is proportional to the applied electric field $\vec{E}$, so that $\vec{V} = \mu\vec{E}$ where the constant of proportionality μ is known as mobility. $\mu_e >> \mu_i$. Mobility is sometimes known as the **forced diffusion.**

diffusion, thermal (or particle) the random flow of heat (or particles) down a thermal (or density) gradient.

dispersion perturbations giving rise to oscillations at one point are propagated with velocities which are called the propagation velocity of the wave. In simple cases this velocity is independent of the frequency. However, in plasma physics we frequently encounter propagation velocities which are frequency dependent. This dependent is called dispersion. in the presence of dispersion, we must differentiate between the phase velocity and group velocity in wave propagation.

dispersion relation it is a relation of the type $\omega^2 = \omega_p^2 + c^2K^2$, where the wave vector $\vec{K}$ is related to the wave frequency ω. It is evident from the relation that as long as $\omega > \omega_p$, K is real. If $\omega < \omega_p$, K is purely imaginary, which means that the waves cannot penetrate into—but are totally reflected from—the plasma.

dispersion, anomalous using the relation $v_g = v_p - \lambda \frac{dv_p}{d\lambda}$, it is noted that in the case in which the phase velocity does not depend on λ, v_p and v_g are identical and the medium is **non-dispersive**. For $dv_p/\lambda > 0$, $v_g < v_p$, which is the case of **normal dispersion**. In the opposite case, $dv_p/d\lambda < 0$, i.e. $v_g > v_p$, the dispersion is said to be anomalous.

disruption, disruptive instability a complex phenomenon involving MHD instability which results in rapid heat loss and termination of a discharge. Plasma control may be lost, triggering a VDE in which the apparatus may be damaged, particularly in large machines. This pheno-menon places a limit on the maximum density, pressure and current in a TOKAMAK. See Hugill diagram.

distribution function let us assume that the macroscopic property of the plasma is due to the average behaviour of innumerable individual particles, where the motion of each individual particle is controlled by the usual rules of mechanics. Suppose there are N_α particles of type α in a given system. For the complete description of the α-particles, we have to know the position and velocity of each particle definitely, but for the statistical description of the plasma, it is necessary only to know the distribution function of the particles.

In the six-dimensional phase-space $(\vec{r}, \vec{V})$ the distribution function of the α-type particle $f_\alpha(\vec{r}, \vec{V}, t)$ is so defined that

$f_\alpha(\vec{r}, \vec{V}, t) d\vec{r} d\vec{V}$ = number of particles of type α at a time t in the volume element $d\vec{r} d\vec{V}$.

DITE divertor and injection TOKAMAK Experiment at Culham. Now dismantled, but experiments that were performed on DITE to investigate disruption control are receiving renewed interest, particularly for Next Step devices.

divergence operator div. in the phase-space it can be written as

$\text{Div}(\vec{V}f) = div_r(\vec{V}f) + div_v(\vec{V}f)$.

divertor a magnetic field configuration affecting the edge of the confinement region, designed to divert impurities/ helium ash to a target chamber. Alternative to using a limiter to define the plasma edge. See H-mode, pumped divertor, radiative divertor, single/double null.

Doublet-III a TOKAMAK run by the General Atomic Company, San Diego, in which the plasma geometry can be varied over a very wide range. See reverse shear. *DIII-D Home Page.*

drift kinetic theory kinetic theory which accurately describes plasma processes which have spatial scales much larger than the particle Larmor radii. See gyrokinetic theory.

drift orbits particle motion is in the direction of the magnetic field, but, particularly for fast particles, electric fields and gradients of the magnetic field give rise to an additional drift perpendicular to the magnetic field.

drift velocity (see diffusion, forced)

driven current plasma current produced by a means external to the plasma (e.g. a transformer, neutral beams, RF waves).

D-T reaction the principal reactions, which involve deuterium (D) and tritium (T) atoms are:

$$D + D \rightarrow {}^3He + n + 3.2 \text{ Mev}$$
$$D + D \rightarrow T + p + 4.0 \text{ Mev}$$
$$D + T \rightarrow {}^4He + n + 17.6 \text{ Mev}$$

The cross-sections for these fusion reactions are appreciable only for incident energies above 5Kev.

dusty plasmas common name for plasmas that, besides electrons and ions, contain macroscopic particles of nanometer to micrometer size. As a rule, such particles are negatively charged. Particles of 10 micron diameter can carry 5,000 elementary charges. By their electrical interaction these highly charged particles can form a liquid or solid state (plasma crystal).

dynamical conductivity in the V-I characteristics of a gas discharge, $\frac{dI}{dV}$, which is a measure of the dynamical conductivity, can be negative. Examples are: metastable states of the glow and arc discharges.

dynamical friction to understand the physical significance of $\langle \partial \vec{V} \rangle$ and $\langle \partial \vec{V} \partial \vec{V} \rangle$ consider the behaviour of a stream of test particles with velocity $\vec{V}$ injected into a plasma. A unit time later, they have slowed down as a result of collisions (in the average) to approach the average field particle velocity. This average change in velocity vector is $\langle \partial \vec{V} \rangle$ and the quantity $\langle m \partial \vec{V} \rangle$, which has the dimension of force, is called **dynamical friction.**

E

$\vec{E} \times \vec{B}$ drift also called the electric field drift $v_{\vec{E}}$ of the guiding centre.

In magnitude, this drift is

$$v_{\vec{E}} = \frac{E(V/m)m}{B(tesla)\sec}$$

It is important to note that $\vec{V}_E$ is independent of q, m and $v\perp$.

earth's atmosphere in the daytime, an appreciable fraction of the solar radiation is absorbed by our atmospheric gases like N_2, C, O_2, CO_2, H_2O etc. Consequently, at a suitable wavelength, this radiation produces photo-ionization. Thus an important region of the atmosphere, known as the ionosphere, is created, the gas in which is a dilute plasma consisting of electrons and ions. the different regions of the ionosphere, namely, D, E_1, E_2, F_1, F_2 layers, are designated in terms of their height, electron number density, temperature etc. For example, the D-layer has an average height of 70 km and electron number density $10^{10}m^{-3}$ approximately, while the F_2-layer has the corresponding values of 300 km and $10^{12}m^{-3}$.

In the field of research in the earth's upper atmosphere, rocket-driven photometers and mass spectrometers have helped much. In the earth's upper atmosphere, a few special types of plasmas exist. **Air glow**—a faint luminescence of the night sky originating in photochemical reactions in the upper atmosphere, also referred to as geocoronal emission—is one of them. **Night glow** during night and—day glow during day are observed, respectively, from the earth's surface and from rockets. **Thin light glow** is another spectacular glow observed during sunset.

A naturally created belt or trap for ions and electrons has been observed in the earth's magnetic field, which is known as the Van Allen Radiation Belt.

echoes, plasma since Landau-damping does not involve collisions or dissipation, it is a reversible process. The truth of this is vividly demonstrated by the remarkable phenomenon of plasma echoes. The experimental set-up is somewhat as follows: A plasma wave with frequency ω_1 and wavelength λ_1 is generated at the first grid and propagated to the right. The wave is Landau-damped to below the threshold of detectibility. A sound wave of ω_2 and λ_2 is generated by a second grid at distance l from the first one. The second wave also damps away. If a third grid connected to a receiver tuned to $\omega = \omega_2 - \omega_1$ is moved along the plasma column, it will find an echo at a distanc $l' = l\omega_2/(\omega_2 - \omega_1)$. What happens is that the resonant particles causing the first wave to damp out retains information about the wave in their distribution function. If the second grid is made to reverse the change in the resonant particle distribution, a wave can be made to reappear. Clearly, this process can occur only in a very nearly collisionless plasma. In fact, the echo amplitude has been used as a sensitive measure of the collision rate.

edge localised mode an instability which occurs in short periodic bursts during the H-mode in divertor TOKAMAKS. It causes transient heat and particle loss into the divertor which can be damaging.

Einstein coefficients the coefficients A and B, known as the Einstein coefficients, are both dependent only on the properties of the atom. The coefficient A is independent of the electromagnetic field and refers to spontaneous emission and the coefficient B refers to the radiation stimulated by the external electromagnetic field.

Einstein relation:

$$\frac{D}{\mu} = \frac{kT}{e}$$

electrodeless discharge if the applied electric field is steady or of low frequency, the discharge phenomena in the tube are nearly the same except that, in the latter case, the extinction potential of the discharge is slightly less than the starting or breakdown potential. But in the case of an h.f. electric field, the mechanisms of the discharge change fundamentally. In studies of the latter in addition to the usual plane electrodes inside the classical discharge tube, external electrodes are also used. The discharge is then called **electrodeless.**

electromagnetic pump in addition to confining plasmas, the magnetic pressure can be used to accelerate plasmas to high velocities. There are a number of methods for projecting individual plasmoids (or plasma blobs) or continuous plasma jets from a nozzle by using the magnetic pressure. As in the confinement problem, a magnetic field can be created by currents—flowing either in external conductors or in the plasma.

The essence of the plasma accelerator process can be understood on the basis of the conducting fluid model. In fact, the same principle is applied in electromagnetic pumping which is used to transport molten metals. The simplest device of this type is the conduction electromagnetic pump (See figure—next column). In this pump, a channel with a conducting fluid is placed between the poles N and S of a magnet. A current I is passed through the fluid at right angles—both to the magnetic field and the channel axis.

An interaction takes place between the mutually perpendicular current and the magnetic field and the pondermotive force drives the fluid stream Q in the direction shown in the figure. The constant current is generated in an external electric field and is applied to the fluid through electrodes. This is a conduction current and is the reason why the pump is called a conduction pump. There are also electromagnetic pumps in which the current is not introduced through the electrodes, but is excited (induced) by an alternating magnetic field. Such pumps are called induction pumps.

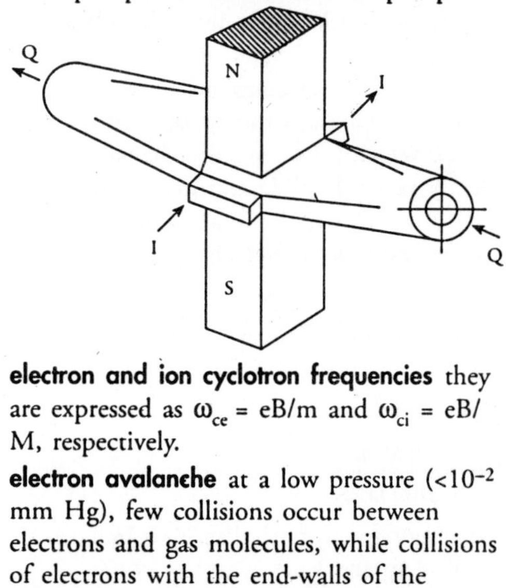

electron and ion cyclotron frequencies they are expressed as $\omega_{ce} = eB/m$ and $\omega_{ci} = eB/M$, respectively.

electron avalanche at a low pressure ($<10^{-2}$ mm Hg), few collisions occur between electrons and gas molecules, while collisions of electrons with the end-walls of the container are frequent. To explain breakdown which occurs in this case at a much lower voltage (~100V), a new mechanism known as the **secondary electron resonance** or **multifactor effect** has been proposed. The basic assumptions of the mechanism are:

(a) transit time of an electron must be equal to or odd multiple of the half-period of the applied electric field so that secondary electrons may be emitted from the electrodes at the proper phase to be accelerated again, and

(b) the secondary electrons acquire sufficient energy from the field so that they can, on falling upon the remote wall, emit at least one electron on the average. Conditions (a) and (b) must be satisfied for an electron avalanche to develop.

electron cyclotron damping this is a strong form of damping which occurs when some particles experience a perturbing force, which oscillates at their cyclotron frequency. Both electron cyclotron damping and ion-cyclotron damping can occur.

In this type of damping, energy is lost, that is, the energy of the electromagnetic field of the wave is converted into the random thermal energy of the particles. This process is used to heat the plasma (the application of this is important in the thermonuclear reaction). In cyclotron damping, the wave energy can be converted into the energy of the particles in a relatively short distance. Since the transfer of energy from electron gas to ion gas is very slow, the ion cyclotron damping is used in most cases, so that the loss of energy by the waves is a straightforward gain of energy by the ions.

electron cyclotron emission radiation emitted by electrons as a result of their cyclotron motion around magnetic field lines. Used to measure electron temperature.

electron cyclotron resonant heating additional heating method using RF waves to accelerate electrons orbiting in the magnetic field. See high field ECRH launch.

electron plasma waves if thermal motion is taken into account, the dispersion relation is given by

$$\omega^2 = \omega_p^2 + k^2 V_{rms}^2$$

where V_{rms} is the root mean square velocity of the particles. Here, then, $d\omega/dk \neq 0$. Therefore such longitudinal vibrations propagate through the plasma by thermal motion and are called **electron plasma waves**. They behave just like sound waves within the plasma.

electrostatic ion cyclotron waves these waves have dispersion relation $\omega^2 = \Omega_c^2 + k^2 v_s^2$, where $\Omega_c = eB_0/M$ is the ion cyclotron frequency. The physical. explanation of these waves is:

The ions undergo an acoustic type oscillation, but the Lorentz force constitutes a new restoring force giving rise to the Ω_c^2 term. The acoustic dispersion relation $\omega^2 = k^2 v_s^2$ is valid if the electrons provide Debye shielding. Electrostatic ion cyclotron waves were first observed by Motley and D'Angelo, in a Q-machine.

electrostatic probe the electrostatic probe (Langmuir probe) can yield the electron temperature from its current-voltage curve. By varying the probe voltage, the Maxwellian electron distribution is sampled. The electrostatic probe was the first plasma diagnostic and is still the simplest and the most localized measurement device. Unfortunately, material electrodes can be inserted only in low-density, cool plasmas.

elongation ratio of the plasma cross-sectional height to its cross-sectional width. See vertical displacement event, plasma geometry.

energy break-even when heating power = total fusion power produced (i.e. alpha power plus neutron power for a D-T plasma). The fusion performance of a power plant is denoted by Q, which is the ratio of the energy of the fusion products to that used to heat the plasma. Break-even corresponds to $Q = 1$, ignition corresponds to $Q = \textit{infinity}$. A burning plasma has $Q > 1$.

energy conservation it is a direct consequence of the collisionless Boatmen equation $df/dt = 0$ that the density of any bunch of points in $(\vec{r}, \vec{V})$ space remains constant in the course of motion. Since the a priori probability of finding a particle in the six-dimensional unit volume is uniform in phase space, the thermodynamic probability of any system of points remains constant as well. This is equivalent to the conservation of entropy of any part of the system.

In thermodynamics, entropy-conserving processes are called adiabatic. In this sense each process involving collisionless plasma in any kind of electromagnetic field (including self-fields) is an adiabatic process.

envelope soliton when an electron plasma wave goes nonlinear, the dominant new effect is that the ponderomotive force of the plasma waves causes the background plasma to move away, causing a local depression in density called a **caviton.** Plasma waves trapped in this cavity then form an isolated structure called an **envelope soliton or envelope solitary wave.** Such solutions are described by the nonlinear Schrödinger equation.

epsilon effect scondary electron emission by excited atoms, in the metastable state, incident on the cathode is called the ε-effect.

error fields the magnetic coils of a TOKAMAK are designed to give the desired magnetic field configuration. Inevitably however, the finite number of coils and imperfections in their construction lead to unwanted deviations from this configuration known as error fields. These can lead to disruptions and are of particular concern for larger TOKAMAKs. See ripple.

eta-effect it is the same as photo-ionization.

extinction potential of the discharge it is the potential at which the discharge stops.

F

Faraday rotation a plane polarized wave sent along a magnetic field in a plasma suffers a rotation of the plane of polarization. This is due to the difference in phase velocity of the R and L waves. In interstellar space, the path lengths are so long that Faraday rotation is important even at very low densities. This effect has been used to explain the polarization of microwave radiation generated by mass action in clouds of OH or H_2O molecules during the formation of new stars.

FER it is a relatively old generation of TOKAMAK, the expanded name being Fusion Experimental Reactor (Japan).

Fermi mechanism if AL is the longitudinal adiabatic invariant, L, the length between a stationary and a moving magnetic mirror, then one obtains, for the approximate invariance of A_L, the energy of a particle trapped between the two mirrors

$= w_z = {}^1/_2 m v_z^2 = mA_L^2/8L^2$.

Thus w_z is inversely proportional to L^2 and increases as L decreases. This method of accelerating a charged particle is called the Fermi mechanism, since it was suggested by Fermi as a possible mechanism for the origin of high energy cosmic rays. The corresponding acceleration is called Fermi acceleration. An alternative theory for the acceleration of cosmic rays is due to Alfvén, who notes that since $\mu = w_\perp/B \approx$ constant, the energy of charged particles can be increased if they simply move in a region of increasing B fields.

Ferraro's law of isorotation consider a stationary, axially symmetric flow of an ideal plasma ($\sigma = \infty$) such that each fluid element moves in a horizontal circle whose centre lies on the vertical Z-axis. Let $\Omega(S, Z)$ be the angular velocity of the particles on the horizontal circle specified by the cylindrical coordinates (S, $0 \le \phi < 2\pi$, Z). The magnetic field is also assumed stationary and axially symmetric, but otherwise unspecified. It is found that Ω must be constant along a magnetic field line $(\vec{B}.\vec{\nabla})\Omega = 0$

This is Ferraro's law of isorotation. It also follows directly from the fact that the magnetic field lines are attached to the plasma.

field coils the coils of a magnetic confinement device that produce the magnetic field required to stabilise and shape the plasma. They may be resistive or superconducting. See TF and PF coils, error fields, cryopump.

field lines, flux surfaces imaginary lines marking the direction of a force field. These define surfaces, to which the particles are approximately constrained, known as flux surfaces. See drift orbits

figure of merit for fusion the condition for a high probability of fusion is the

Lawson criterion:

$n\tau \geq 1.5 \times 10^{22}$ with n in particles per cubic metre and a τ in seconds. It is clear that both high density and a long confinement time are of equal importance; hence the product $n\tau$ is often used as a figure of merit for fusion devices.

FIR far Infra-Red (e.g. wavelength about 0.2-1 mm) part of the electromagnetic spectrum. FIR lasers are used to measure the magnetic field and plasma density.

fire hose instability it is found that in a plasma of electrons and a single kind of ions, the Alfvén waves are unstable if

$P_{11} > p\perp + (B_0^2)/4\pi$.

The reason is that a small bulge in the magnetic field lines has a tendency to grow owing to the centrifugal force exerted by the particles that move along the field lines. This instability is, therefore, approximately called the **fire hose or garden hose instability**. The above criterion can be roughly understood by noting that p_{11} is a measure for the particle velocity along the field line, $B_0^2/4\pi$ represents the magnetic force tending to straighten the field line and $p\perp$ is an additional stiffness due to the particles gyrating around the field line.

first order orbit theory for the study of the motion of plasma particles in magnetic and electric fields, two approaches are adopted, namely, the microscopic or individual and macroscopic or collective. In the former approach, the nature of the trajectory or the orbit of each particle, under external or internal electric and magnetic fields, is considered. The theory developed with this approach is known as the **first order orbit theory**, since here collisions and other interactions among charged particles are neglected as a first order approximation. Here, collisions are regarded as a small perturbation, in contrast with an ordinary gas, where collisions play an important role. This approach is valid for a low density and high temperature plasma. For the physical point of view, this theory is beyond doubt and quite useful, because much information about the behaviour of the plasma can be obtained with its help. Nevertheless, it has its limitations and is not of much use in the complex problems where orbits of many particles have to be considered.

The macroscopic approach, on the other hand, is a more realistic one in that it takes into account the particle interaction which always takes place in real plasma.

fishbones rapid bursts of MHD activity sometimes observed when neutral beam heating is used in TOKAMAKs (fishbone refers to the shape of the bursts in magnetic field when plotted as a function of time).

Fick's law Fick's law of diffusion is given by

$\vec{\Gamma} = -D\vec{\nabla}n$

This equation merely expresses the fact that diffusion is a random walk process, in which the net flux from dense region to less dense region occurs simply because more particles start in the dense region. This flux is obviously proportional to the gradient of the density. in plasmas, Fick's law is not necessarily obeyed. Because of the possibility of organized motions (plasma waves), a plasma may spread out in a manner which is not truly random.

fluorescent tube low pressure gas discharge in a mixture of the rare gas argon and mercury vapour. In the positive column of this discharge, the mercury atom is excited by electron impact to emit ultraviolet radiation. A layer of luminescent material on the inner tube wall converts the UV-radiation to visible light.

flute instability a simple kind of hydromagnetic instability occurs when a plasma, with no internal field, comes in contact with a magnetic field in free space. If the strength of the magnetic field decreases with distance from the boundary of the plasma (in which case the field lines must be convex outward), then it is energetically favorable for the plasma to change places with the magnetic field. This phenomenon is called the interchange instability. Because the motion can be described as an interchange of the lines of force in space. In this case, the field lines move as a whole, conserving their form and direction. The surface perturbation takes the shape of flutes along the magnetic field lines. Therefore, this instability is also known as a flute instability.

flux conservation the magnetic flux passing through a surface is defined as

flux = $\int \vec{B}.d\vec{S}$.

Thus for a particle spiraling about the field lines

$$\text{Flux} = B\pi r^2 = B\pi \frac{m^2 V^2 \perp}{e^2 B^2} = \frac{2\pi m W\perp}{e^2 B} = \frac{2\pi m}{e^2}\mu$$

Thus the conservation of W⊥/B or μ may be thought of as a conservation of the flux passing through the cyclotron orbit of the particle. Also it can be noted that since Br^2 is a constant, an increase in B leads to a decrease in the cyclotron radius.

flux equation for electrons the electron flux $\vec{\Gamma}$ can be written as $\vec{\Gamma} = n\vec{u} = -De\vec{\nabla}n$. This equation shows that the electron flux is caused by density gradients.

flux loop a large, single-turn coil of wire, used to measure the magnetic flux enclosed by it.

Fokker-Planck code a computer code to describe the velocity distribution of plasma particles allowing for collisional relaxation and plasma heating. Calculates distribution functions.

Fokker-Planck equation in the calculation of binary elastic collisions, an important term known as the Boltzmann's collision integral is obtained, which, when expanded, gives the Fokker-Planck equation. Both of them are of much importance towards understanding of properties of the plasma. Derivation of the Fokker-Planck equation is long and complex. the derivation of the equation is generally given by two separate methods—one of them is the Markoff process and the other is the test particle theory.

fourth state of matter plasma is sometimes defined as the fourth state of matter, resembling gaseous state at high temperature.

free-electron diffusion coefficient from Einstein's relation D/μ = kT/e, we have $D = \frac{\mu kT}{e} = \frac{e}{mV}\frac{kT}{e} = \frac{kT}{mV}$. This is called the free electron diffusion coefficient, μ being electron mobility and ν being collision frequency.

Fresnel's formulas the description of plasma in terms of a refractive index is not quite correct. There are two reasons why Fresnel's formulas for reflection and refraction do not readily apply. The **first reason** is that these formulas presuppose a sharp boundary, i.e., the refractive index must drop from its value in the medium to its vacuum value at a distance which is short compared to the wavelength. This is not true for most plasmas but it is true for electrons in a metal or semiconductor.

frozen-in-field lines the magnetic lines of force are, in essence, permanently tied to the fluid particles, both move together with the same velocity $\vec{V}_D$. This is known as the frozen-in-field (particles move with the lines of force, the lines of force move with the particles). Due to collisions, however, particles may get shifted from one line of force to another resulting in the diffusion of plasma across the magnetic field.

fuelling supplying a fusion plasma with the necessary reactants. See pellet injection.

full wave theory wave theory which includes complete accounting of wave energy (transmitted, reflected and absorbed, including energy transferred to other waves) for studying RF heating.

fusion a fusion reaction occurs when two light nuclei (ions) approach each other so closely that their Coulomb (charge) repulsion is overcome, allowing the nuclei to fuse. The total mass of the fusion products is lower than that of the two original nuclei; the difference is converted to kinetic energy which is distributed between the products. Methods being investiga
an attempt to harness this potentially huge source of energy include magnetic

confinement fusion and inertial confinement fusion. See Lawson Criterion, deuterium-tritium reaction, deuterium-helium-3 reaction.

fusion power core the main components of a fusion power device, which for a TOKAMAK include the limiter or first wall, divertor, blanket, shield, PF coils, TF coils, current drive system, cryostats and structural components. Items such as buildings, turbines, and heat exchangers etc., which are not specifically fusion-related, are not included.

fusion product the product of a fusion reaction, for example an alpha-particle or neutron in a deuterium-tritium plasma.

fusion reactivity fusion reaction rate per ion. for present typical TOKAMAK conditions, it increases with the density and temperature of the plasma.

fusion triple product mathematical product of density, temperature and energy confinement time. A measure of approach to energy break-even and ignition. See also Lawson Criterion.

G

g.c. (guiding centre) approximation the orbit of a particle initially moving perpendicularly to $\vec{B}$ is a circle. The centre of this circle is known as the **guiding centre.** The magnetic moment $\mu = W\perp/B$ is an adiabatic invariant. in the limit of infinitely slow variation of $\vec{B}$, namely, $\Delta\vec{B}$ small in space over several Larmor radii and in time over several Larmor periods, we can take $\partial B/\partial t \approx \Delta B/T_c$ and arrive at the result $\mu = W\perp/B$ = constant. This approximation made here is known as the **guiding centre** or **G.C. approximation.**

Γ (gamma) space it is the same as phase space. It is the space needed to describe the system in the micro-canonical ensemble, since from Γ space we may determine the total energy or other constants of the motion directly.

gas discharge in glass tubes filled with a gas at low pressure (below 10 mbar) an electric current can flow at moderate applied voltages (500-1,000 V), which causes intense luminous phenomena. such gas discharges are widely used e.g. as *neon tubes*. See also: fluorescent tube.

gas laser it is a plasma device used to produce high-intensity coherent light-signals.

generalized Ohm's law

the equation $\vec{E} + \vec{V} \times \vec{B} = \eta\vec{j}$

is called the generalized Ohm's law, η being the D.C. resistivity

$\eta = 1/\sigma = m\nu/ne^2$; when $\nu \to 0$, $\sigma \to \infty$, that is, the d.c. conductivity of a plasma is infinity in the absence of collisions; thus, for a collisionless plasma, we must have, for infinite $\vec{j}$, $\vec{E} + \vec{V} \times \vec{B} = 0$; this equation is usually known as Ohm's law for that plasma, the d.c. conductivity of which is infinity.

Gibbs distribution function it is a well-known result of statistical mechanics that in thermodynamic equilibrium a solution of the Lionville equation is the Gibbs distribution function.

Gibbs-Boltzmann law the ion and electron density distributions according to Gibbs-Boltzmann law are:

$n_i = n_{i0}e^{-e\phi/KT_i}$

$n_e = n_{e0}e^{-e\phi/KT_e}$

where T_i, T_e represent the ion and electron temperatures, K is Boltzmann constant and n_{i0}, n_{e0} are the average number densities of ions and electrons, in the absence of the test particle. In a plasma, $n_{i0} = n_{e0} = n_0$ (say).

glow, night (see Earth's atmosphere)

glow, twilight (see Earth's atmosphere)

glow discharge in the region beyond Townsend discharge (See dark discharge), the V-I characteristics show a peculiar nature. The current continues to increase even if the voltage is reduced. This region, thus, constitute a metastable state when dV/dI, and, hence, the dynamic conductivity is negative. It is a transition zone between the Townsend discharge and glow discharge that immediately follows. In the glow discharge region, current continues to increase at constant voltage and the value of V and I are, respectively, about 10^2 volts and 10^{-4}volts to 10^{-2} amp. The glow discharge has been studied most extensively.

glow, air (see Earth's atmosphere)

glow, day (see Earth's atmosphere)

glow, (negative) originally part of a low pressure gas discharge close to the cathode. Because of similar mechanisms, those parts of high frequency discharges are also called glow that are sustained by energetic electrons from the boundary layers.

Grad B drift it is given by

$\vec{V}_{\nabla B} = \frac{V\perp^2}{2\omega_c} \frac{\vec{B} \times \vec{\nabla}\perp B}{B^2}$ (symbols have usual meanings).

It can also be expressed as

$$\vec{V}_{\nabla B} = \frac{mV\perp^2}{2eB^3} \vec{B} \times \vec{\nabla}\perp B \; (\because \omega_c = \frac{eB}{m})$$

Grad-Shafranov equation a second-order, non-linear elliptic partial differential equation that describes the relationship between the plasma current distribution and pressure in terms of magnetic flux. This equation is readily solved numerically by computer, and the calculated plasma geometry and profiles that satisfy it comprise a plasma equilibrium

gravitational current-density

$$\vec{j}_g = e(n_i \vec{V}_{gr} - n_e \vec{V}_{ge}) = ne(\vec{V}_{gi} - \vec{V}_{ge})$$

It is given by $= \frac{M+m}{eB^2} ne(\vec{g} \times \vec{B}) = \frac{\sigma}{B^2}(\vec{g} \times \vec{B})$

where $\sigma = (M + m)n \approx Mn$ = mass density of the charged particles.

gravitational drift gravitational drift is given by $\vec{V}_g = \frac{m}{eB^2}(\vec{g} \times \vec{B})$

In this drift motion, the drift for ions and electrons are in opposite directions and are directly proportional to their masses.

gravitational drift velocity the drift velocity in uniform gravitational and magnetic fields is given by $\vec{V}_g = \frac{m}{eB^2}(\vec{g} \times \vec{B})$

gravitational instability in plasma, Rayleigh-Taylor instability can occur because the magnetic field acts as a light fluid supporting a heavy fluid (plasma). In curved magnetic fields, the centrifugal force on the plasma due to particle motion along the curved lines of force acts as an equivalent "gravitational" force. That is why this type of instability is called gravitational instability.

Greenwald density the Greenwald normalised density is given by $n_{20}.pi.a^2/I_p$, where n_{20} is the electron density expressed in units of $10^{20}m^{-3}$, a is the plasma minor radius in metres, and I_p is the plasma current in mega amperes. In many TOKAMAKs this value does not exceed 1, so the Greenwald density is a measure of the density limit for a TOKAMAK.

guiding centre plasma in the description of plasma it is sometimes sufficient to follow the motion of the guiding centre of the particles. this is often called **guiding centre plasma.**

gyrofrequency (see cyclotron angular frequency)

gyro-kinetic theory version of kinetic theory in which the radius of Larmor gyration is not assumed to be small. An essential theory for investigating fine-scale instabilities which might be responsible for driving turbulence, which may 'in turn' be responsible for anomalous transport.

gyro-relaxation effect the constancy of the diamagnetic moment μ can be utilized to heat plasma in the trap to high temperature by means of an alternating magnetic field. Resonance occurs when the field frequency approaches the cyclotron frequency; the particles can then extract large amounts of energy from the field. This is known as the **gyro-relaxation effect.**

gyrotron device used for generating high power microwaves in the electron cyclotron range of frequencies (about 50-200 GHz).

H

hall effect in deriving generalized Ohm's law, one comes across a term $\vec{j} \times \vec{B}$, which

is often negligible but not always. This term is called the Hall current term.
The occurrence of the Hall Effect is explained as:
when a current carrying conductor in the form of a bar is placed in a magnetic field at right angles to the current, a voltage is developed in the conductor in a direction at right angles to both the current and the magnetic field. This phenomenon is known as the Hall Effect and the voltage is called the Hall voltage.
In the steady state, the force on the charge carriers due to the Hall field will balance the magnetic field. We then have
$qE_H = qvB$, where the symbols have their usual meanings.

$\therefore E_H = vB = \frac{J}{nq}B$ (J = current density = nqv;
n = concentration of the carriers.)
E_H is the hall field,
$1/nq$ = **hall coefficien** R_H (say).
$\therefore E_H = R_H BJ$
$\therefore$ hall coefficient R_H may be defined as the Hall field produced by unit magnetic induction when the current density in the bar is unity.

Hall effect in mhd problems the approximations made with regard to the generalized Ohm's law lead to the simple form $\vec{J} = \sigma(\vec{E} + \vec{V} \times \vec{B})$(see generalized Ohm's law). In this derivation, the term involving $\vec{J} \times \vec{B}$ is retained, which leads to a Hall effect in mhd flow problems.

halo current current which flows outside the confined plasma region, in the scrape-off layer. During a VDE, the plasma makes contact with a limiter and starts to be scraped off, causing a fraction of the plasma current to flow along the field lines. The current path can intercept a material surface, where the current will take the path of least resistance, causing intense heating and deformation of the material. The effects of halo currents may not be axisymmetric, and can cause great damage. They are a potential concern for Next Step devices.

hamilton's canonical equations the equations are: $\vec{p}_i = -\frac{\partial H}{\partial \vec{q}_i}$ and $\vec{q}_i = \frac{\partial H}{\partial \vec{p}_i}$
where $\vec{p}_i$ and $\vec{q}_i$ are generalized momenta and coordinates, respectively, and $H(\vec{p}, \vec{q}, t)$ is the Hamiltonian function.

Hartmann number plasma flow in a magnetic field (magnetohydrodynamic flow) is similar to the flow of a fluid or gas in that it can be either laminar, i.e. in smooth layers, or turbulent, i.e. vortex flow. In most cases, a magnetic field directed along the flow inhibits the formation of turbulence, that is to say, it stabilises the laminar motion. The stabilizing action of the magnetic field is strongest in a highly conducting fluid, in which the field-freezing effect applies.
The nature of the flow of a nonconducting fluid or gas is determined by the dimensionless Reynolds number
$R_e = vL/\nu$, where v is the flow velocity, L is a linear dimension and ν is the kinematic viscosity. At small Reynolds numbers, the flow is laminar; at large numbers, it becomes turbulent. In magnetohydrodynamics, the character of the motion depends on more than one dimensionless number. If we replace the kinematic viscosity by the magnetic viscosity $1/\sigma\mu_0$ in the Reynolds number we obtain the magnetic Reynolds number:
$R_{e_M} = \sigma\mu_0 vL$
in the case of large magnetic Reynolds numbers, the characteristics of the motion are determined by the ratio of the magnetic pressure $B^2/2\mu_0$ to the pressure head $\sigma V^2/2$:

$$A = \frac{B^2/_{2\mu_0}}{\rho V^2/_2} = \frac{B^2}{\mu_0 \rho V^2}$$

For large values of the parameter A (greater than 0.1) the magnetic field completely stabilises the laminar flow of a high conductive plasma. For small magnetic Reynolds numbers the character of the flow is determined by the dimensionless Stuart number

$$S = AR_{e_M} = \frac{B^2}{\mu_0 \rho V^2}.\sigma\mu_0 vL = \frac{B^2 \sigma L}{\rho V}$$

All the criteria given above apply to flows along the magnetic field. The nature of the flow across a magnetic field in a channel of width L is determined by the value of the dimensionless Hartmann number

$$M = \sqrt{SR_e} = \sqrt{\frac{B^2 \sigma L}{\rho V} \frac{\nu L}{\nu}} = BL\sqrt{\frac{\sigma}{\rho V}}$$

For small values of M, the flow is the same as with no magnetic field.

heat flux the heat flux $\vec{Q}$ is defined as the excess thermal energy per unit area per unit time and is given by $\vec{Q} = -K\vec{\nabla}T$ where K = thermal conductivity.

heat flux vector flow vector or heat flux vector is given by

$$\vec{Q} = \frac{1}{2} m_\alpha \int \omega_\alpha{}^2 \vec{W}_\alpha f_\alpha d\vec{V}$$

where random velocity in excess of the average velocity of α-type particles $\vec{W}_\alpha = \vec{V} - \vec{u}_\alpha$, $\vec{V}$ = actual (random) velocity of any particle, u_α = average velocity of all the α-type particles and wa = $|\vec{W}_\alpha|$.

helias a stellarator configuration in which the coils resemble distorted, non-planar TF coils—no continuous helical coils or TOKAMAK-like PF coils are present. The Helias (HELIcal Advanced Stellarator) has been proposed to be the most promising stellarator concept for a power plant, with a modular engineering design and optimised plasma, MHD and magnetic field properties. The Wendelstein VII-X device is based on a five field-period Helias configuration.

helicity injection a means of imparting a helical magnetic wave into a plasma to produce a driven current and is potentially valuable for sustaining toroidal and spherical devices. The helicity of a toroidal plasma is related to a linkage of toroidal and poloidal magnetic fields, and is approximately conserved throughout a discharge. If additional helicity can be injected, the plasma current can be sustained or even increased.

helicons these are waves of the type of Whistlers in the electron gas of a metal. Whistlers are produced by electrical discharges in the atmosphere (thunderstorms), which excite electromagnetic waves in the audiofrequency range. These waves propagate mainly along the field lines of the earth's magnetic field, so that they re-enter the atmosphere at a point on the opposite hemisphere. Whistlers are received as short a whistle, whose pitch are high at the beginning and subsequently drops. The same whistler may be received several times because it is reflected back and forth along the field lines.

heliotron a stellarator configuration in which a helical coil is used to confine the plasma, together with a pair of PF coils to provide a vertical field. TF coils are sometimes also used to control the magnetic surface characteristics.

heliotron-E a large stellarator of heliotron configuration, built in Kyoto, Japan.

helium ash fusion reactions in a deuterium-tritium plasma produce energetic alpha-particles (helium nuclei), which heat the plasma as they slow down. Once this has happened, the alpha-particles have no further use: they constitute helium ash, whose removal and replacement by deuterium-tritium fuel is required to prevent dilution of the plasma. See argon frosting.

helium burning the initial stage of evolution of a star is **hydrogen burning**, in which hydrogen burns, by two different processes, into helium. The next phase of evolution of stars is **helium burning**. When hydrogen is used up as a nuclear fuel, the star contracts and gain energy from its own gravity. When He^4 is used up, the next phase is **carbon burning**.

helix, pitch of the pitch p of a helix is given by

$$p = V_{11} \cdot \frac{2\pi}{\omega_c} = 2\pi V_{11} \frac{m}{eB}$$

Clearly, the pitch of an ion is much greater than that of an electron, p is less if V_{11} is less; then the orbit will look like a compressed spring. Similar will be the appearance of the orbit if B is higher.

H factor ratio of the energy confinement time for a given set of plasma conditions to the theoretical value calculated from an L-mode scaling law. for a plasma in the H-mode the H-factor is typically around 2, that is, the confinement time is enhanced by a factor of two compared to the normal L-mode.

hierarchy of moment equations in order to determine the values of n unknowns, we need n equations. But we find that the n-th moment equation requires knowing n+1 quantities. Unless this hierarchy of moment equations is truncated at some stage, a finite closed set of equations for the macroscopic quantities can never be obtained.

high beta condition in which the plasma energy is a significant fraction of the energy in the magnetic field. an alternative measure is when the plasma energy is comparable to the energy in the poloidal magnetic field (i.e. poloidal beta is approximately unity). See beta.

high field ECRH launch electron cyclotron resonant microwaves launched from the inside of the plasma torus. This allows higher density plasma to be heated.

homogeneous distribution the distribution function is said to be homogeneous when it is independent of the position co-ordinate $\vec{r}$.

homopolar experimental device rotating plasmas are interesting from certain viewpoints. According to theory, rotation of plasma can help achieve stable plasma configuration. Using rotating plasmas, certain plasma traps—namely, the Ixion and Homopolar experimental devices—have been built.

"Hot" plasma a plasma of which the temperature ≈ a few hundred eV and for which electrical conductivity σ is very high (in contrast, "cold" plasma has temperature ≈ a few eV and electrical conductivity σ is low. Also, collision results in leaking of the magnetic field through the plasma)

H-regime (mode) a High confinement regime that has been observed in TOKAMAK plasmas. It develops when a TOKAMAK plasma is heated above a characteristic power threshold, which increases with density, magnetic field and machine size. It is characterised by a sharp temperature gradient near the edge (resulting in an edge "temperature pedestal"), ELMs, and about a 100% increase in energy confinement time compared to the normal L-regime.

H-theorem Boltzmann's equation

$$\frac{\partial f}{\partial t} + \vec{V}.\vec{\nabla}_r f + \frac{f_e}{m}\vec{\nabla}_w f = \left(\frac{\partial f}{\partial t}\right)_{coll}$$

possesses several interesting properties:

(i) there are some functions of $(\vec{V})$ for which $\int\left(\frac{\partial f}{\partial t}\right)_{coll} \psi d\vec{V} = 0.$

(ii) if $f(\vec{r},\vec{V}, t = 0)$ is initially positive, then it is always positive.

(iii) this equation possesses the properties of irreversibility, called the H-theorem.

(iv) this equation leads the system towards some equilibrium stage, that is, any initial distribution of particle velocities will change with time through collisional processes and the final distribution will be an equilibrium distribution.

For a homogenous gas, Boltzmann's H-theorem states that the quantity

$$H = \int f \ln f d\vec{r} d\vec{V}$$

known as the Boltzmann H-function, obeys the relation $\frac{\partial H}{\partial t} \le 0$

The equilibrium condition is determined by

$$\frac{\partial H}{\partial t} = 0$$

Hugill diagram a plot of $1/q$ against nR/B (where q is the safety factor at the edge of

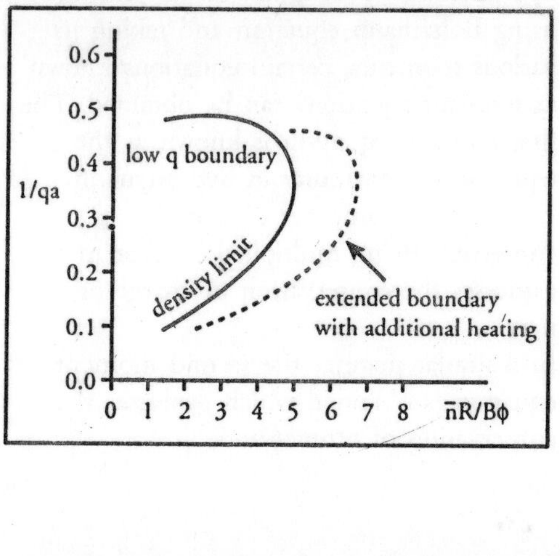

the plasma, n is the average electron density, R is the plasma major radius and B is the toroidal field at the plasma centre), showing the boundary of TOKAMAK operation as limited by disruptions. The density and q limits are clearly seen. (In the following Hugill diagram n is in units of 10^{19} electrons per cubic metre.)

hybrid frequencies wave-propagating at right angles to the magnetic field exhibit two frequencies at which anomalous dispersion occurs. These frequencies do not, in general, coincide with the cyclotron frequencies. these anomalous dispersion frequencies are called hybrid frequencies.

hydrodynamic approximation in the conducting-fluid model a number of forces which act on the plasma are related to the pressure force. In an ordinary gas of neutral particles, pressure transfer takes place solely by means of collisions. Thus, the pressure force acts only at densities high enough so that frequent collisions occur. An approximate method for describing equilibrium and gas motion in terms of a pressure is called the **hydrodynamic approximation.** The hydrodynamic approximation can also be applied to dense plasmas just as for any dense gas. However, even in tenuous plasmas in which collisions do not play a significant role, one can imagine a pressure that acts on each particle at right angles to the magnetic field.

A description of the behaviour of tenuous plasmas in terms of a pressure is called the quasi-hydrodynamic approximation.

hydrodynamic description of the plasma using Boltzmann equation and taking its various moments, certain equations known as moment equations can be obtained. The first moment equation is known as the equation of continuity in hydrodynamics. It states that the number of particles is conserved. Being multiplied by e or m it expresses the conservation of charge or mass.

In a similar manner the second moment equation is obtained which expresses the conservation of momentum.

The first two moment equations are truncated from the rest by assuming that the pressure tensor is a scalar, that is, all its non-diagonal terms vanish and the diagonal terms equal to the ordinary scalar pressure p(= nKT), so that

$$\vec{\nabla}.\vec{P} = \vec{\nabla}p$$

For any particular problem, these two moment equations are solved along with Maxwell's equations. This approach valid for a collision dominated (and hence in local equilibrium) plasma is known as the hydrodynamic description of the plasma.

hydrodynamic model in the hydrodynamic model, plasma is assumed to be an interacting gas mixture, in which electron gas, one or more types of ion gas and neutral molecular gas having number densities n_e, n_i (= n_e) and n_0, respectively, co-exist amicably, each species, however, retaining its own set of equations for mass- momentum and energy-transfer.

hydromagnetic shock wave the wave magnetic field $\vec{B}(t)$ is time-dependent and, hence it will sometimes be less, sometimes greater. Hence, the velocity of larger portions of the wave will be greater than that of smaller portions. As a result, waves of large amplitude will overtake those of small amplitude, that is, steepening will occur, which will be converted to a type of wave known as **hydrodynamic shock wave.**

hydromagnetic instabilities instabilities play a central role in plasma phenomenon. In many cases, the equilibrium of a plasma is stationary (by 'stationary' we mean a condition which does not change with time unless it is subjected to a perturbation. Stationarity should not be confused with instability), i.e. in the absence of a perturbation the plasma state can persist indefinitely, although it is unstable to an arbitrarily small perturbation.

I

IAEA International Atomic Energy Agency

ICE ion cyclotron emission. Analogous to ECE. Observed in JET and TFTR as a suprathermal signal, apparently driven by collective instability of energetic ion populations such as fusion products and injected beam ions.

ideal in the context of magnetohydrodynamics, "ideal" implies that the magnetic field and the plasma always move together. For this to occur, the electrical resistivity of the plasma must be negligible.

ideal internal kink modes a magnetohydrodynamic instability of the central region of a TOKAMAK. This, or its close relative the resistive internal kink mode, may be involved in the sawtooth disruptions which occur in most TOKAMAKs.

ignition condition condition for self-sustaining fusion reactions: for a deuterium-tritium plasma, ignition occurs when the heating of the plasma by the alpha-particles is equal to the heat loss. See energy break-even.

impact parameter an electron of velocity $\vec{V}$ approaches a fixed ion of charge e. In the absence of Coulomb forces the electron would have a distance of closet approach γ_0 called the impact parameter.

inertial fusion common name for experiments for fusion with lasers and ion beams. In both cases energy is deposited in such a short time that, because of its inertia, the target sphere keeps together sufficiently long to yield the required number of fusion reactions.

initial value problem the response of the plasma to an initial perturbation can be studied as an initial value problem for solution of Vlasov equation. Lauden's analysis of the process gives a clear physical picture of Landau damping. One of the most important results of this analysis is that, in general, there is not a single frequency associated with each wavelength, but rather a number of possible frequencies. Hence, there is no dispersion relation in the usual sense. Most of these frequencies are complex and, hence, the oscillations of the potential either damp in time or grow.

instabilities, classification of in the classification of plasma waves, we assume an unperturbed state which is one of perfect thermodynamic equilibrium. The particles have Maxwellian velocity distribution and the density and magnetic field are uniform. In such a state of highest entropy, there is no free energy available to excite waves and we have to consider waves that are excited by external means. We now consider states that are not in perfect thermodynamic equilibrium. The free energy which is available can cause waves to be self-excited, the equilibrium is then an unstable one, giving rise to an instability. Instabilities may be classified according to the type of free energy available to drive them. There are four main categories:

(a) beam-plasma instability a cold electron beam of a certain density δn_0, say, and velocity u shot into a cold plasma of density n_0 at rest, gives rise to a beam-plasma instability. Assume that an electron beam penetrates a plasma and the velocity of the electrons in the beam is considerably higher than the thermal velocity of the plasma electrons, while the density N_b of the electrons in the beam is considerably lower than the density N_0 of the plasma electrons. Deceleration of the electron beam can occur owing to the scattering of the electrons of the beam on the electrons and ions of the plasma. There is, however, another mechanism of deceleration of the electron beam, which is known as the **Langmuir paradox** or beam instability. This kind of deceleration can be more effective than deceleration due to scattering on the charged plasma particles. This mechanism acts as follows. The plasma oscillations are generated in plasma. Interacting with the electrons of the beam and taking energy from them, these oscillations are amplified. Thus the energy of the electron beam is transformed into the energy of plasma oscillations and it remains in the plasma.

(b) Buneman instability the natural frequency of oscillations in the electron fluid and the natural frequency of oscillations in the ion fluid can coincide in the laboratory frame because of the Doppler shift in the moving electron fluid. The density fluctuation of ions and electrons can then satisfy Poisson's equation. The total kinetic energy of the electrons is less when the oscillation is present than when it is absent. The electron oscillations have negative energy and the ion oscillations have positive energy. Both waves can grow together while keeping the total energy of the system constant. Instability of this type is used in klystrons to generate microwaves.

(c) **drift instability** it is a universal instability also called resistive drift wave instability.

(d) **fire-hose or garden hose instability** in a plasma of electrons and a single kind of ions, the Alfvén waves are unstable if $p_{11} > p\perp + (B_0)^2/4\pi$.

The reason is that a small bulge in the magnetic field lines has a tendency to grow, owing to the centrifugal force exerted by the particles that move along the field lines. This instability is, therefore, appropriately called **the fire hose or garden hose instability**. The above criterion can be roughly understood by noting that p_{11} is a measure for the particle velocity along the field line, $B^2/4\pi$ represents the magnetic force tending to straighten the field line, and $\phi\perp$ is an additional stiffness due to the particles gyrating around the field line.

(e) **the gravitational instability** (See Rayleigh-Taylor instability)
in plasma, Rayleigh-Taylor instability can occur because the magnetic field acts as a light fluid supporting a heavy fluid (the plasma). In curved magnetic fields, the centrifugal force on the plasma due to particle motion along the curved lines of force is an equivalent gravitational force.

(f) **interchange instability** consider two thin adjacent flux tubes in the fluid.
A possible perturbation is one that interchanges these tubes without disturbing the rest of the system. If this lowers the potential energy of the system, it must be unstable.

(g) **kinetic instabilities** in fluid theory, the velocity distributions are assumed to be Maxwellian. If the distributions are, in fact, not Maxwellian, there is a deviation from thermodynamic equilibrium; and instabilities can be derived by the anisotropy of the velocity distribution. For instance, if T_{11} and $T\perp$ are different, an instability called the modified Harris instability can arise. In mirror devices, there is a deficit of particles with large $v_{11}/v\perp$ because of the loss cone; this anisotropy gives rise to "loss cone instability".

(h) **Rayleigh-Taylor instabilities** in this case, the plasma has a density gradient or a sharp boundary, so that it is not uniform. In addition, an external non-electromagnetic force is applied to the plasma. It is this force which drives the instability. this happens whenever a heavy fluid is supported by a light fluid, as is well-known in hydrodynamics.

(i) **streaming instabilities:** in this case, either a beam of energetic particles travels through the plasma, or a current is driven through the plasma, so that the different species have drifts relative to one another. The drift energy is used to excite waves, and oscillation energy is gained at the expense of the drift energy in the unperturbed state.

(j) **universal instabilities** even when there are no obvious driving forces as an electric or a gravitational field, plasma is not in perfect thermodynamic equilibrium as long as it is confined. The plasma pressure tends to make the plasma expand, and the expansion energy can drive instability. This type of free energy is always present in any finite plasma and the resulting waves are called universal instabilities.

(k) **velocity space instabilities** not all instabilities are equally dangerous for plasma confinement. A high frequency instability near ω_p cannot affect the motion of heavy ions. Low frequency instabilities with $\omega << \Omega$, however, can cause ambipolar losses via $\vec{E} \times \vec{B}$ drifts. Instabilities with $\omega \approx \Omega_c$ do not efficiently transport particles across $\vec{B}$ but are dangerous in mirror machines, where particles are lost by diffusion in velocity space into the loss cone.

(l) mirror instability instability of the type related to the fire hose instability, is the mirror instability. The criterion that the plasma suffers from mirror instability is

$$p_{\|} < \frac{p\perp^2}{p\perp + (B_0)^2/8\pi}$$

(m) instability, Kink (m = 1), (more popular: screw-shaped instability). The magnetic field is clearly stronger on the inside than on the outside, so that the kink continues to grow. Again, an axial magnetic field improves the stability, because the tension along the field lines tries to straighten the kink.
For higher m, one also finds unstable eigen values but these are even more easily made stable by the applied field in the z-direction.

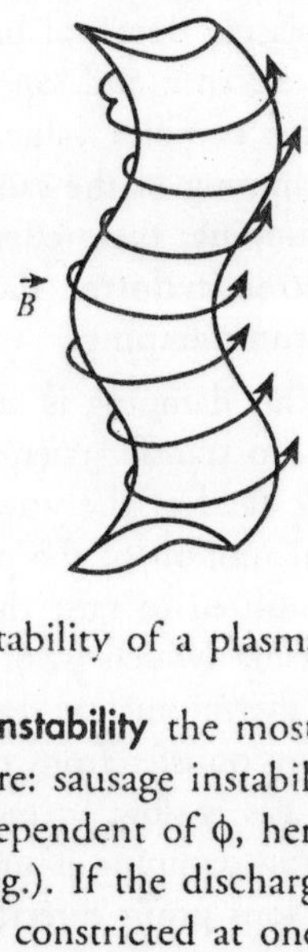

Kink instability of a plasma column

(n) sausage instability the most important instabilities are: sausage instability (m = 0). independent of ϕ, hence axially symmetric (fig.). If the discharge happens to be slightly constricted at one place, the external magnetic field increases at that place and hence also the magnetic pressure. Therefore the constriction continues to increase. This instability can be reduced by applying an additional axial magnetic field within the plasma. An applied magnetic field in the z-direction outside the plasma also helps to prevent the sausage instability, because its field lines are stretched by a constriction of the discharge, which amounts to an increased magnetic energy.

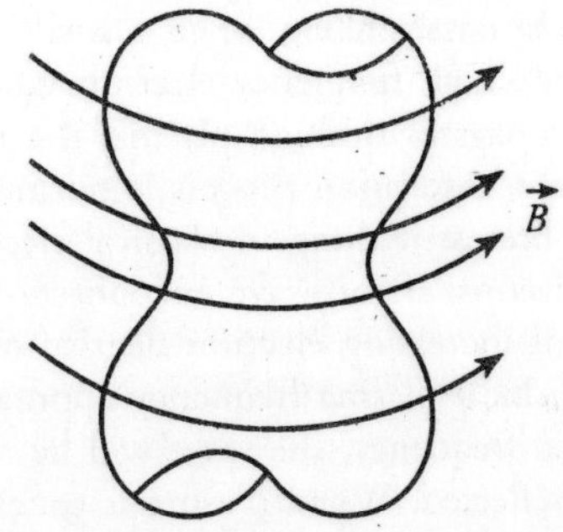

Sausage instability of a plasma column

(o) two-stream instability when plasma becomes unstable for long wave longitudinal oscillations, it is called the "two-stream instability". Plasma of electrons and ions is unstable when the electrical current density is too high. This is another form of two-stream instability.

INTOR abbreviation for International Tokamak Reactor.

invariant adiabatic

the first adiabatic invariant $\mu = \frac{W_\perp}{B} = \frac{mV_\perp^2}{2B}$

It is invariant in spatially and temporally varying $\vec{B}$ fields. The period motion involved is the Larmor gyration.

the second adiabatic invariant J also called the longitudinal invariant J and is defined for a half-cycle between the two turning points a and b:

$J = \int_b^a v_{11} ds$

It can be proved that J is invariant in a static nonuniform B field; the result is also true for a slowly time-varying B field.

the third adiabatic invariant the slow drift of a guiding centre around the Earth constitutes a third type of periodic motion. The adiabatic invariant connected with this turns out to be the total magnetic flux ϕ enclosed by the drift surface. As B varies, the particle will stay in a surface such that the total number of lines of force enclosed remains constant. This invariance, ϕ, has few applications because most fluctuations of $\vec{B}$ occurs on a time scale short compared with the drift period.

Inverse bremsstrahlung when a small amplitude high frequency electromagnetic wave propagates through plasma, the most important absorption process is normally inverse bremsstrehlung in classical electron-ion collisions. If the wave encounters a region of increasing electron density such that the local plasma frequency approaches the wave frequency, the wave will be almost totally reflected. When plasma is generated by focusing a laser beam into a solid target, the density gradient will usually be extremely steep, resulting in little penetration, ion absorption and only a small fraction of the incident energy transferred to the plasma.
If the amplitude of the wave is increased sufficiently, the interaction with the plasma becomes strongly non-linear and several new effects are theoretically predicted. The rate of energy absorption by inverse bremsstrahlung reaches a maximum when the velocity of electrons oscillating in the radiation field equals their thermal velocity.

inverse pinch when a high current passes along the axis of plasma, an electrodynamic force is produced. This compresses the cross-section of the plasma. In this category, there are also **inverse pinches**, where the plasma is driven outward by a current flowing through a conductor in the centre of the plasma.

ion acoustic shock waves when a jet travels faster than sound, it creates a shock wave. This is a basically nonlinear phenomenon, since there is no period when the wave is small and growing. The jet is faster than the speed of waves in air, so the undisturbed medium cannot be "warned" by precursor signals before the large shock wave hits it. In hydrodynamic shock waves, collisions are dominant. Shock waves also exist in plasmas, even when there are no collisions.

ion acoustic wave the expression for the velocity C_s of sound waves in a neutral gas is:

$$C_s = \frac{\omega}{k} = \left(\frac{\Gamma p_0}{\rho_0}\right)^{\frac{1}{2}}$$

The waves are pressure waves propagating from one layer to the next by collisions among the air molecules. In plasma with no neutrals and few collisions, an analogous phenomenon occurs. This is called an **ion** acoustic wave or, simply, an ion wave.

ion Bernstein wave a wave which only exists in a hot plasma and is supported by the ions. It propagates at right angles to the magnetic field, when it is undamped, at harmonics of the ion cyclotron frequency. There is also an electron Bernstein wave which propagates at harmonics of the electron cyclotron frequency. See magneto-acoustic cyclotron instability.

ion cyclotron current drive non-inductive current drive using ICRH.

ion cyclotron damping when the refractive index of the plasma does not become infinite at the electron and ion resonances, but takes a large complex value, it corresponds to a damping of the same kind as the Landau damping; these effects are called the **electron cyclotron damping** and the **ion cyclotron damping**.

The effect of this damping is to dissipate energy, that is, to transfer energy from the electromagnetic field of the wave to the disordered heat motion of the particles. This effect is utilized to raise the temperature of the plasma (which is, of course, important for thermonuclear reactions). Since the energy transfer from the electron gas to the ion gas is slow, in most cases, the ion cyclotron damping is utilized in order that the ions profit directly from the dissipated energy.

ion cyclotron resonant heating additional heating method using RF waves at frequencies (about 20-50 MHz) matching the frequency at which ions gyrate around the magnetic field lines.

ionosphere uppermost layer of the atmosphere, between 60 km and 500 km altitude, which is ionised by ultraviolet radiation from the Sun. The existence of the ionosphere was deduced from the reflection of radio waves, which allows transatlantic short-wave communication. Since the ionosphere represents an electrically conducting plasma.

isotope atomic nuclei with same number of protons but different number of neutrons are called isotopes.

ITER abbreviation for International Thermonuclear Experimental Reactor.

ixion a plasma trap where rotating plasmas are used to achieve stable plasma configuration.

J

JET abbreviation of Joint European Torus (England).

Joule heat if we imagine a swinging pendulum which is free to oscillate without being disturbed, the oscillation decays gradually because of friction and the oscillation energy is transformed into heat. This process is an example of energy dissipation. Plasma oscillations also decay because of various dissipative processes. The simplest mechanism for dissipation in plasmas is collisions between particles; these transform the energy of ordered oscillation into heat, i.e., into random particle motion. If an electric current flows in a plasma, as a result of the electrical resistance, i.e., finite plasma conductivity, part of the energy of the current is transferred into heat. This heat, liberated by the current, is called Joule heat. Joule dissipation is a special case of dissipation due to collisions since the electrical resistance of the plasma results from collisions between the current-carrying particles and other particles in the plasma.

JT-60U JT-60U is the flagship TOKAMAK of the Japanese magnetic confinement programme, similar in size to JET. It is sited at Naka and run by JAERI. A much larger TOKAMAK is planned—JT-60SU. See reverse shear. *JT60-U* Home Page.

Jupiter some information has been received about Jupiter in recent years. Particularly, the following reaction has been proposed for the structure of the ionosphere of Jupiter:

$He^{+} + H_2 \rightarrow HeH^{+} + H \rightarrow He + H_2^{+}$.

K

kinematic viscosity (ν) given by $\eta = \rho\nu$, where η = coefficient of viscosity, ρ = density.

kinetic equation the same as Boltzmann equation.

kinetic theory a detailed mathematical model of a plasma in which trajectories of constituent electrons and ions are described. More complex than fluid and two-fluid theories, it is necessary in the study of RF heating and some instabilities, particularly when energetic particles are involved. See drift and gyro kinetic theories.

kink instability (See instability, kink)

Kortewig-de Vries equation this equation occurs in many physical situations including that of a weakly nonlinear ion wave.

Kruskal-Schwarzschild instability the dispersion relation ω^2 = kg leads to unstable solutions for every wave number, as expected (since ω is imaginary). In fact, the larger is k (the smaller the wave length) the faster the instability grows. This is Kruskal-Schwarzschild instability.

L

Landau damping it is a damping process named after Let Landau (1908-1968), who first derived it. it is a **collisionless** process, that is, the velocity distribution itself may be used to show addition or subtraction of energy from the wave, depending upon the sign of the slope of the velocity distribution function at the phase velocity ω/k.

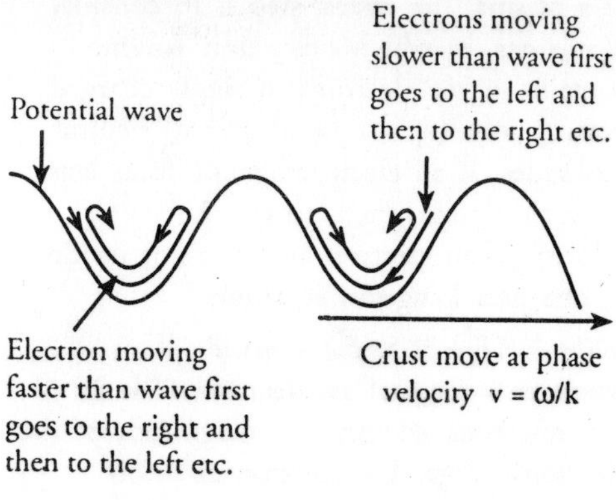

the physical mechanism for such damping may be looked at in the following way: As a result, traveling potential waves are set up. These potential waves tend to "trap" particles in them. Those particles whose velocities are exactly equal to the wave phase velocity are, of course, trapped automatically. Those particles whose velocities are slightly greater than the wave phase velocity must attempt to surmount a wave potential hill. If the particle does not have enough energy to go over the top of the hill, it will "roll back" down the hill, up the other side, oscillate between the two crusts, and, eventually, be trapped. Since the particle had been moving faster than the wave, it has now been decelerated, on the average, and therefore, has lost some of its energy to the wave. The wave energy is thus increased and grows in amplitude because of the gain in energy obtained from this particle. However, at the same time, those particles moving slightly slower than the wave phase velocity are speeded up, that is, they get some more energy from the wave, which results in a damping effect on the wave. In the region of velocity space where the distribution function has a slope that is negative, there are slightly more particles moving at the speed that produces damping than at the speed that produces growth of the wave. The net result is that when the slope of the distribution function is negative at $v = \omega/k$, damping occurs. When the slope is positive at this point, then growth of the wave amplitude occurs.

Langevin equation to know the properties of a plasma, the easiest step is to consider it as a gas, in which every unit volume contains n free electrons, n singly charged positive ions ($n_e \approx n_i = n$) and n_0 neutral molecules. If an electromagnetic force acts on such plasma, then, to get the average velocity of an electron at any point inside it, one uses Langevin equation:

$$m\vec{u}' = -e(\vec{E} + \vec{u} \times \vec{B}) - m\nu\vec{u}$$

where m = mass of an electron, $\vec{u}$ is its instantaneous velocity, e = magnitude of the electron charge, ν = electron collision frequency, that is, number of collisions per second of the average electron with the heavier particles of the plasma (ions and neutral atoms), $\vec{E}$, $\vec{B}$ electric and magnetic fields acting on the electron.

Langmuir probe of the various diagnostic techniques, the most versatile one is the Langmuir probe, which provides measurements of electron density and temperature and of ion density. The probe is a small metal electrode (usually cylindrical or spherical, though sometimes in the shape of a disk); the essence of the technique is to monitor the current to the probe when different voltages are applied.

A typical probe characteristic displaying the variation of probe current with voltage is shown in the figure.

Is V becomes less negative with respect to the plasma, some of the energetic electrons from the tail of the electron distribution are collected by the probe until, eventually, the electron current drawn just cancels the ion current. This occurs when V = Vp , which is known as the floating potential. A further increase in V leads to a rapid increase in electron current until this saturates at space potential V_s on account of the space-charge limiting of electron current. From the figure, $V_s \approx 0$.

It can be shown that the logarithm of the electron current plotted as the function of probe potential is linear and the slope of the line allows the electron temperature to be determined.

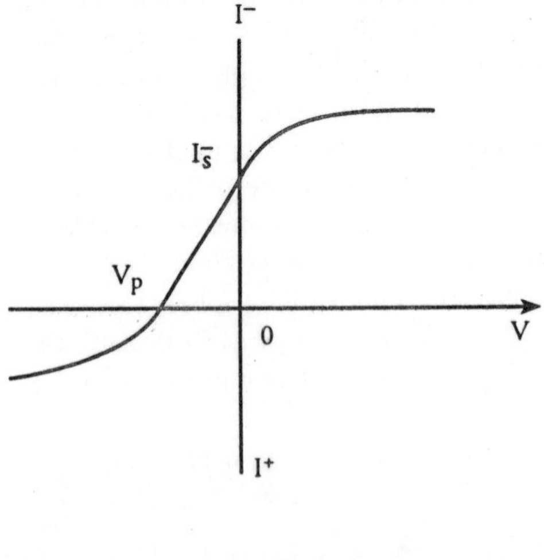

Having determined the electron temperature, one can obtain the electron density. One can also determine the ion density in the plasma. The ion temperature, on the other hand, is not obtainable from Langmuir probe measurements.

The Langmuir probe can provide information in the presence of a magnetic field, but interpretation of the characteristics is less straight-forward.

Various modifications of the Langmuir probe exist, such as the double probe technique. A probe of another kind is the radio-frequency plasma probe developed by Sayers—primarily for satellite measurements of ionospheric and exospheric electron density.

large helical device the world's largest stellarator, situated in Nagoya, Japan, at the National Institute for Fusion Studies. LHD started operating in March 1998. *LHD Home Page.*

larmor radius radius of the gyratory motion of particles around magnetic field lines.

laser ablation use of lasers for disintegration and ionisation of objects injected into the plasma.

laser fusion a scheme for nuclear fusion, where compression and heating of a bead made of deuterium and tritium (called the target) is accomplished by irradiation of intense laser light.

laser-plasma interaction (see Inverse Bremsstrahlung).

Lauson criterion (see confinement).

levitron in connection with the Z-pinch, complete stability is predicted if B_θ falls off everywhere in the plasma at least as rapidly as 1/r. We should, therefore, expect stability if B_θ is due to a current along the axis. This is, essentially, the basis for certain plasma experimental devices termed 'hardcore' and 'Levitron', the latter being the toroidal counterpart of hardcore cylindrical devices, and possessing a magnetically levitated central ring.

In this device, a magnetic field configuration of high shear is obtained in a toroidal vacuum region by producing toroidal and poloidal components of the magnetic field. The whole system of conductors is cooled to superconducting temperatures. An additional set of toroidal coils produces a vertical field, to create a magnetic well between the ring and the major axis. With this configuration, a complete stability is obtained.

L-H transition change from L-mode to H-mode (usually quite sudden).

LHe / ScHe liquid helium or supercritical helium used in fusion to cool superconducting magnets and cryopumps.

lightning electrical discharge at high gas pressure, which occurs between thundercloud and soil. The electric current in the lightning channel can reach values of 10,000 A. The plasma and the neutral gas inside the channel is strongly heated by the current flow. This causes a sudden expansion of the lightning channel which can be heard as thunder. It could only recently be proved that lightning discharges occur between cloud and ionosphere.

limiter a material surface within the TOKAMAK vessel which defines the edge of the plasma and, thus, avoids contact between the plasma and the vessel. Alternative to using a divertor to define the edge.

line radiation radiation due to the decay from excited states by electrons. Excitation of line radiation arises from electron impact and from recombination into excited states.

linear pinch (also called Z-pinch). When a high current passes along the axis of a plasma, an electromagnetic force ($\vec{j} \times \vec{B}$ force) is produced which compresses the cross-section of the plasma.

Several types of pinch devices have been developed. Among them, the Z-pinch and the θ pinch should be mentioned. In the Z-pinch, the current density along Z is J_z and the magnetic field in the azimuthal direction θ is B_θ; the ($\vec{j} \times \vec{B}$) force is directed towards the axis. On the other hand, if the external magnetic field is applied along the longitudinal direction and if its magnitude is greater than the magnetic field internally produced by the current in the plasma, then the plasma is converted into a pinch, being compressed by this magnetic field. This process of compression of the plasma is also used in some controlled nuclear fusion devices. The conversion of the plasma into a pinch due to compression by the external magnetic field is known as the θ pinch. Devices based on θ pinch configuration are known as **Thetatrons.** Note that $\vec{j}$, $\vec{B}$ lines of the linear and θ pinches are interchanged. That is, the plasma of the Z-pinch is confined by the azimuthal current.

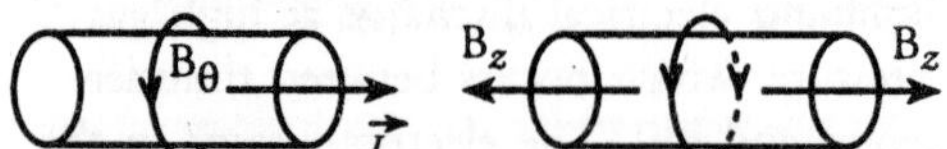

Z-Pinch θ-Pinch

Liouville equation it states that in the six-dimensional coordinate-velocity space or phase space particle number-density is conserved, if the observer moves with the phase-points in the phase space.
By taking moments of the Liouville equation with respect to the velocity variables, conservations of several quantities e.g. particle number, momentum and energy, are obtained, as in the case of Boltzmann's equation.

LN2 liquid nitrogen used to protect LHe circuits from heat ingress and in cryopumps to trap water vapour or carbon dioxide that may leak into the vacuum vessel.

Lorentz force the Lorentz force on a charged particle is given by
$\vec{F} = \theta(\vec{E} + \vec{V} \times \vec{B})$

Lorentz gas an electron gas is a stationary background of heavy particles, which has just enough uniformly distributed positive charge density to make the plasma macroscopically neutral when the plasma is undisturbed.

Loschmidt's number it is the number density of an ideal gas at 1 atm and 273 K.

loss cone with reference to the magnetic mirror, the particles which are lost from the magnetic bottle are those within the solid angle σ (fig.). (Note that this figure refers to velocity space and not configuration space). This solid angle σ defines the loss cone. The probability P, of loss from the bottle, is

$$P = \frac{\sigma}{2\pi} = \int_0^{\theta} \sin\theta d\theta$$

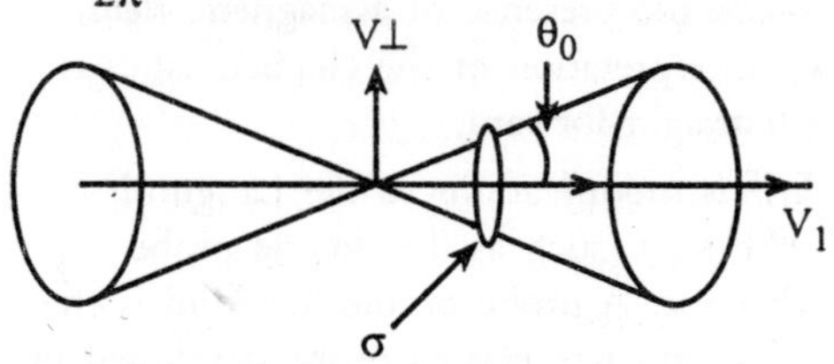

Loss cone for a magnetic bottle.
Those particles with resultant velocities lying within the cone will penetrate the magnetic mirror and be lost from the bottle.

lower hybrid current drive non-inductive current drive using lower hybrid heating.

M

Mach number it is the ratio of the particle velocity to the local sound velocity.

magnetic (diamagnetic) moment μ = W⊥/B, where W⊥ is the transverse kinetic energy, B is the magnetic field.

magnetic axis the magnetic surfaces of a TOKAMAK form a series of nested tori, of decreasing minor radius. The central "torus" defines the magnetic axis.

magnetic confinement fusion the use of strong magnetic fields to confine plasma so as to allow fusion reactions to occur within it. See TOKAMAK, stellarator, reverse field pinch, Z-pinch, theta-pinch.

magnetic diagnostics coils of wire in various configurations, placed around the outside of a plasma to measure the local magnetic field or related quantities. See flux loop, Mirnov coil, Rogowski coil.

magnetic double-mirror and trap in the special arrangement of magnetic field such that $B = B_0$ at the middle and $B = B_{max}$ at the two ends. B_0 the lowest and B_{max} is the highest. Under certain conditions, the particles will be trapped, being reflected again and again between the two ends.

Confined between magnetic double-mirror, this arrangement of oscillatory motion of the particles along the lines of force is known as the **magnetic trap or magnetic (double) mirror**. This is also called a **magnetic bottle** or a **magnetic plug**.

magnetic heating it is also called adiabatic compression in which the energy of the particles is increased by slowly increasing a magnetic field in the plasma.

magnetic islands islands in the magnetic field structure caused either by externally applied fields or internally by unstable current or pressure gradients. See tearing magnetic islands.

magnetic mirror using the constancy of magnetic moment μ and the principle of conservation of energy, it can be shown that a converging magnetic field can reflect particles and can behave as a magnetic mirror.

magnetic pressure it is equal to $B_0^2/2\mu_0$.

magnetic probe it is a wire loop introduced into the plasma. (Fig). A changing magnetic field excites a current in the circular loop and this current is measured on an oscillo-scope. Knowing the distribution of the magnetic field and using the laws of electrodynamics, one can calculate the distribution of currents in the plasma.

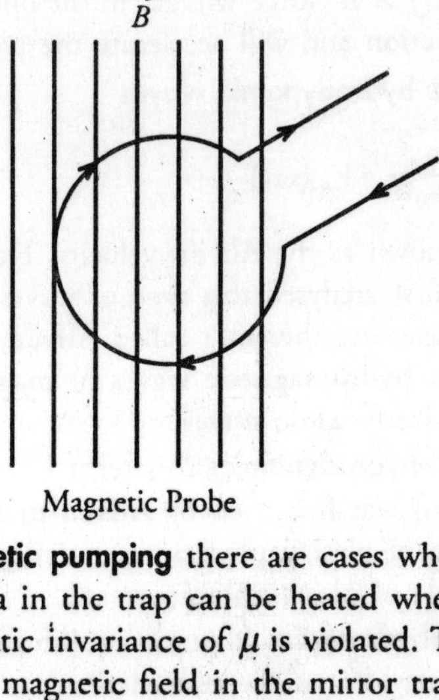

Magnetic Probe

magnetic pumping there are cases where the plasma in the trap can be heated when the adiabatic invariance of μ is violated. Thus, if the magnetic field in the mirror trap is periodic, the transverse velocity $\vec{V}\perp$ would oscillate, though without any gain of energy due to this. However, if collisions occur, adiabatic invariance of μ will not hold and plasma will be heated. Specially, an oscillat-ing particle during its compression phase can transfer a fraction of its gyration energy into longitudinal V_{11} energy. This process is called **magnetic pumping**.

magnetic Reynolds number the nature of the flow of a nonconducting fluid or gas is determined by the dimensionless Reynolds number

$$R_e = \frac{vL}{\eta}$$

where v is the flow velocity, L is a linear dimension and η is the kinematic viscosity. At small Reynolds numbers, the flow is laminar; at large numbers it becomes turbulent. In magnetohydrodynamics, the character of the motion depends on more than one dimensionless number. If we replace the kinematic viscosity by the magnetic viscosity $1/\sigma\mu_0$ in the Reynolds number: $R_{e_M} = \sigma\mu_0 vL$

If the Reynolds number is large, the flow medium can be assumed to be infinitely conducting; if R_{e_M} is small, the conductivity is low.

magnetic scalar potential let $\vec{B} = \vec{\nabla}\phi_m$ with ϕ_m the magnetic scalar potential; while, for the divergence condition, we have

$$\nabla^2\phi_m = \vec{\nabla}.\vec{B} = 0$$

Thus the magnetic field in which the particle moves is not completely arbitrary but must, in fact, be the gradient of a solution to Laplace's equation ($\nabla^2\phi_m = 0$).

magnetic shear a method of avoiding interchange instabilities is based on the requirement that, to be interchanged, the flux tubes need to be adjacent. By intro-ducing shear into the magnetic field, two field lines adjacent in a small region are far removed elsewhere, which prevents their interchange by an infinitesimal pertur-bation. Such is the case, e.g., in the stellarator configurations, where the pitch of the field line spirals varies with the axial distance.

magnetic storm on the earth, magnetic storms are caused by shock waves generated in the interplanetary plasma by a solar flare.

magnetic viscosity the quantity $1/\sigma\mu_0$ (the magnetic diffusion coefficient) is sometimes called the **magnetic viscosity**.

magnetic beach the effect of Landau damping is to dissipate energy, that is, to transfer energy from the electromagnetic field of the wave to the disordered heat motion of the particles. This effect is utilized to raise the temperature of the plasma which is, of course, important for thermonuclear reactions. Since the energy transfer from the electron gas to the ion gas is slow, in most cases the ion cyclotron damping is utilized in order that the ions profit directly from the dissipated energy.

An experimental way to achieve this is the following: B_0 is made not entirely homogeneous, but slowly decreasing in strength in the Z-direction. as a consequence the plasma density slowly increases. One now generates electromagnetic waves in the low density region, whose frequency corresponds to the ion cyclotron frequency of the high density region. These waves travel along the Z-direction into the high density region, where resonance occurs, and their energy is dissipated. This device is called **magnetic beach**.

magneto-acoustic cyclotron instability this instability results from an exchange of energy between the fast Alfvén wave (or magneto-acoustic wave) and an ion Bernstein wave which has a source of free energy through the presence of a population of energetic (non-thermal) ions, e.g. fusion products. The instability occurs for propagation perpendicular to the equilibrium magnetic field. See super Alfvénic velocity.

magneto-acoustic waves these waves can be produced when particle velocity $\vec{V}$ is parallel to the wave vector $\vec{K}$ and both $\vec{V}$ and $\vec{K}$ are perpendicular to the applied magnetic field $\vec{B}_a$ (Fig.). The waves are longitudinal and known as **magnetoacoustic** or **magneto-sonic waves**. The phase velocity of this wave $V_{ph} = (V_S^2 + V_A^2)^{1/2}$ where V_S = velocity of sound and V_A = Alfvén velocity. When $\vec{B}_a \rightarrow 0$, this wave is converted into ordinary sound wave.

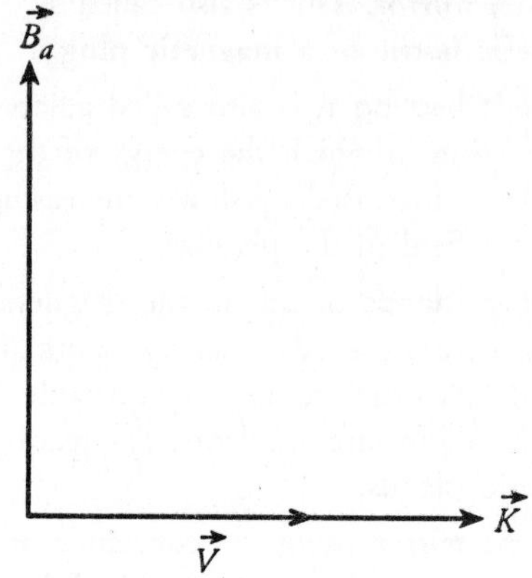

magnetohydrodynamic generator an important possible application of plasma physics is the direct conversion of thermal energy into electric energy with the help of the magnetohydrodynamic generator. The basic principle of this device is very simple. Here much of the energy of the initial thermal energy store of the plasma is converted into kinetic energy by making it unidirectional. The plasma jet formed thereby is inserted perpendicular to the magnetic field.

magnetohydrodynamic motor this device is opposite to magnetohydrodynamic generator. The $\vec{j} \times \vec{B}$ force will act in the opposite direction and will accelerate the plasma.

magneto-hydrodynamic waves

$$V_{ph} = \sqrt{\frac{B_a^2}{\mu_0 \sigma}} = V_A \text{ (say)}$$

V_A is known as the Alfvén velocity. Hannes Alfvén first analysed this type of waves. these waves are therefore called **Alfvén waves** or **hydromagnetic waves** or **magneto-hydrodynamic waves**.

magneto-hydrodynamics this term (M.H.D.) was first used by Alfvén to mean a branch of plasma physics or sometimes the whole of plasma physics.

magnetohydrostatics in static equilibrium, the velocity $\vec{V}$ and the time derivatives are zero and the set of equations of magnetohydrodynamica reduce to the magnetohydrostatic equations:

$\vec{J} \times \vec{B} = \vec{\nabla} p$, $\vec{\nabla} \times \vec{H} = \vec{J}$, $\vec{\nabla}.\vec{B} = 0$.

Now, $\vec{J} \times \vec{B}$ is perpendicular to both $\vec{J}$ and $\vec{B}$. Hence, $\vec{\nabla} p$ is also perpendicular to both $\vec{J}$ and $\vec{B}$ (since $\vec{B}.\vec{\nabla} p = 0$ and $\vec{J}.\vec{\nabla} p = 0$). Now, since $\vec{\nabla} p$ is perpendicular to the isobaric surface it follows that both $\vec{J}$ and $\vec{B}$ lie on the isobaric surface; conversely, pressure is constant along current and lines of force, i.e. p = constant. Surfaces are both "magnetic surfaces" and "current surfaces".

magneto-plasma when a plasma is subject to a strong magnetic field, it is called a magneto-plasma. Its properties are then basically changed, because electrons and ions cannot move freely perpendicular to the magnetic lines of force. The path of each particle then becomes helical, the axis of the helix being parallel to the magnetic field and, in that state, the plasma becomes highly anisotropic.

magnetosphere solar wind compresses the Earth's magnetic field B until the solar wind pressure is equal to $B^2/2\mu_0$. Consequently, the Earth with its magnetic field and Van Allen radiation belt is restricted inside a cavity, known as the magnetosphere, outside which is the solar atmosphere.

mobility the drift velocity $\vec{V}$ is proportional to the applied electric field $\vec{E}$, so that, $\vec{V} = \mu\vec{E}$ where the constant of proportionality μ is known as **mobility**. In the M.K.S. system, μ is expressed in m^2/Volt sec. Mobility of electrons is much greater than that of ions. Mobility is sometimes known as the **forced diffusion.**

major radius distance from the centre of the torus (to the centre of the plasma cross-section). See plasma geometry.

MAST Mega Amp Spherical Tokamak, currently operating at Culham. MAST follows on from the pioneering START device. *MAST Home Page.*

MHD and other plasma instabilities unstable distortions of the shape of the plasma/magnetic field system. See Alfven gap mode, ballooning instability, disruption, edge localised mode, ideal internal kink modes, internal kink, internal reconnection event, kinetic instability, large scale ideal modes, magneto-acoustic cyclotron instability, marfe, microinstabilities, mode number, neo-classical tearing modes, peeling mode, profile control, resistive ballooning modes, resistive instability, resonant magnetic perturbation, sawtooth, TAE modes, tearing modes, vertical displacement event.

microinstabilities instabilities with characteristic lengths similar to the particle Larmor radii, rather than to the TOKAMAK dimensions. These are thought to be responsible for the fine-scale turbulence in TOKAMAKs, and, hence, anomalous transport.

minor radius half the horizontal extent of the plasma cross-section. In the context of plasma profiles, the radial distance from the centre of the plasma. See plasma geometry.

Mirnov coil (pickup coil) a small cross-section, multiple-turn coil of wire, used to measure the component of the local magnetic field perpendicular to the plane of the coil. These coils are generally referred to as "pickup" coils when the output voltage is integrated—this is proportional to the magnetic field. Fluctuations in the field are best measured by using the unintegrated signal—the coils are then referred to as "Mirnov" coils.

mirror ratio the ratio B_{max}/B_0 is known as the **mirror** ratio. If the mirror ratio increases, the loss cone of the magnetic mirror becomes narrower, causing more particles to be trapped.

mobility tensor the definition of mobility e.g. $\vec{V} = \mu\vec{E}$ (see diffusion, forced) can be given in the tensor form $\vec{V} = \overleftrightarrow{\mu}\vec{E}$, where $\overleftrightarrow{\mu}$ is the mobility tensor. In the rotating system, $\overleftrightarrow{\mu}$ is diagonal and we have

$$\begin{bmatrix} v_L \\ v_R \\ v_{11} \end{bmatrix} = \frac{q}{m} i \begin{bmatrix} \frac{1}{\omega_c - \omega} & 0 & 0 \\ 0 & -\frac{1}{\omega_c + \omega} & 0 \\ 0 & 0 & -\frac{1}{\omega} \end{bmatrix} \begin{bmatrix} E_L \\ E_R \\ E_{11} \end{bmatrix}$$

In the stationary Cartesian system with the magnetic field pointing in the Z direction, it reads

$$\begin{bmatrix} v_x \\ v_y \\ v_z \end{bmatrix} = \frac{q}{m} \begin{bmatrix} \frac{i\omega}{\omega_c^2 - \omega^2} & \frac{\omega_c}{\omega_c^2 - \omega^2} & 0 \\ -\frac{\omega_c}{\omega_c^2 - \omega^2} & \frac{i\omega}{\omega_c^2 - \omega^2} & 0 \\ 0 & 0 & -\frac{i}{\omega} \end{bmatrix} \begin{bmatrix} E_x \\ E_y \\ E_z \end{bmatrix}$$

mode another word for wave or oscillation in a plasma. Also used, independently, in place of the word regime.

mode number characterises the wavelength of an instability.

N

Navier-Stokes equation it describes momen tum transport and belongs to hydrodynamics.

NET Next European Torus.

neutral beam a beam of high velocity neutral atoms injected into the plasma to impart momentum to the plasma ions. Neutral beam injection is a method of providing additional heating and current drive. See charge exchange recombination spectroscopy, energetic particle, fishbones, plasma rotation.

neural network a computer code or electronic circuit that uses incoming data to give plasma parameters, etc., having previously been "trained" on a series of examples by analogy with the learning processes involved in the human brain.

neutron elementary particle carrying no electric charge, which was discovered by Chadwick in 1932. (Symbol: n; mass = $1{,}67.10^{-27}$ kg).

neutron stars also called pulsars; they emit radiation in pulses (so the name), causing disturbances in the earth's atmosphere.

new comb criterion this criterion (not given here) is necessary and sufficient for stability and is a non-local criterion only expressible as an abstract mathematical theorem.

non-linear wave equation two solitons (or solitary waves) propagating with different velocities interact when they overlap non-linearly. Then they part company, each one preserving its original shape and velocity. A non-linear wave equation (known as the kortewig de Vries or kdV equation) describes the evolution of the solitary wave.

nonlinear dynamics field of research, which originated from theoretical mechanics. Since about 1980, common aspects of similar complex phenomena and their mathematical description are studied in a multi-disciplinary approach between physics, mathematics, biology, chemistry, control engineering, social sciences and economics.

nonneutral plasmas systems of electrons confined in Paul or Penning traps have much in common with ordinary plasmas. Because of their similar collective effects, such systems are called non-neutral plasmas.

northern light luminous phenomena in the upper atmosphere (ionosphere), which are caused by energetic electrons from the solar wind. At polar latitudes electrons can enter along magnetic field lines and stimulate a glow discharge hervorrufen.

NSTX a spherical TOKAMAK currently operating at Princeton, USA. NSTX is a similar size to MAST, but of different design—the two devices will run complementary experimental programmes. *NSTX Home Page.*

nuclear fusion the fusion of light atomic nuclei sets the binding energy free. For this purpose, the atomic nuclei, despite their electric repulsion, must be brought so close together that they nearly touch. This requires an energy of several MeV (million electron volt). Nuclear fusion therefore a occursat very high temperature insert and in a plasma state.

nuclear reactions in stellar plasmas although the basic principles of nuclear reactions in laboratory and stellar plasmas are the same, some differences exist:

(i) in stellar plasmas, the ion velocity distribution can always be assumed to be Maxwellian, but not in the laboratory plasmas.

(ii) Stellar plasmas are usually much denser than the laboratory plasmas.

nucleon common name for particles constituting the atomic nucleus, i.e. proton and neutron.

number density one defines a function of position, velocity and time, $f(\vec{r}, \vec{V}, t)$ such that $fd\vec{r}d\vec{V}$ is the probability of finding particles within the σ-dimensional volume element $d\vec{r}d\vec{V}$, centered at the point $(\vec{r}, \vec{V})$ in coordinate and velocity space. Observable properties of the plasma can then be obtained from this function, known as the distribution function, by taking various velocity moments of f. The number density, for example, of particles in the spatial volume element $d\vec{r}$ is $n(\vec{r}, t) = \int f(\vec{r}, \vec{V}, t)d\vec{V}$, is where the integration taken over the whole of velocity space.

Nyquist diagram these are curves which were first used to discuss instabilities in electric circuits.

O

ohmic heating (OH) inductive heating by using a transformer to drive a current in the plasma. This is necessarily pulsed.

optimised shear adjusting the current profile to optimise TOKAMAK performance (e.g. see reverse shear).

operating limits see TOKAMAK operating boundaries.

P

parameter space in the CMA diagram, a co-ordinate system is used involving plasma parameters (electron and ion density, magnitude of) as well as mode frequency. This space is known as the parameter space and is divided into a number of volumes by surfaces representing the cut-offs and principal resonances.

parametric instabilities wave-wave interactions giving rise to these instabilities.

Paschen's law in the theory of the Townsend discharge, Paschen's law is written as $V_b = f(pd)$, where V_b is the breakdown voltage, p = gas pressure, d = distance between the electrodes and V_b plotted against pd shows a minimum.

peeling mode an edge MHD instability which exists when the current density at the plasma edge is non-zero. It may be associated with ELMs.

pellet bead of frozen deuterium, which is injected as solid matter into a TOKAMAK plasma for refueling.

pellet injection fuelling a plasma by firing into it a stream of small pellets of frozen material (e.g. deuterium) at high velocity.

Penning effect it can be represented by $A^* + B \rightarrow A + B^+ + e$. Atom A^* is in a metastable state, its excitation energy is higher than the ionization potential of atom B.

Penrose criterion it is a stability criterion for electrostatic plasma oscillations.

PF coils the coils that produce the external poloidal field in a TOKAMAK (in addition to that produced by the plasma current), used to control the plasma geometry.

PINI Positive Ion Neutral Injector—the main component of a neutral beam injection system.

photo-ionization it can be represented by $A + h\nu \rightarrow A^+ + e^-$.

photo-dissociation it can be represented by $AB + h\nu \rightarrow A + B$.

photon quantum of electromagnetic radiation. Important for ionisation processes are photons of visible light (650nm-400nm), ultraviolet radiation (400nm-100nm) and x-rays (<100nm).

pinch effect an interaction between magnetic field and plasma results in the so-

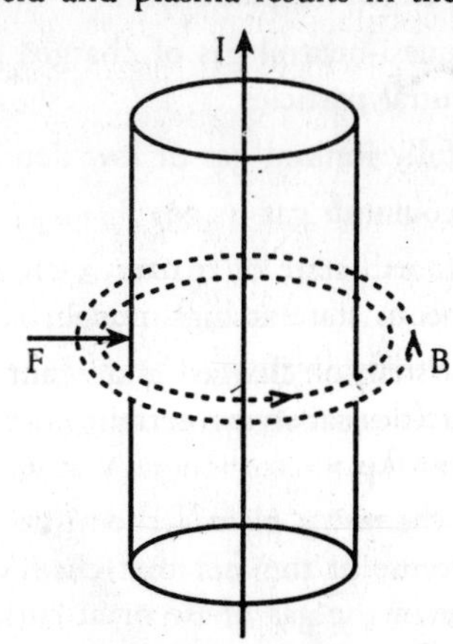

called pinch effect. It occurs at high current discharges. The pinch means more or less a cylindrical tube of plasma confined by a magnetic field. The magnetic field itself is produced by the current flowing through the plasma and this current, again, is produced by the applied magnetic field and its gradient. Therefore, the entire structure is **self-consistent**.

In gas discharges, the current flow produces on the ionized gas an electrodynamic force $\vec{j} \times \vec{B}$ per unit volume. Since this force always acts perpendicularly to the direction of the current, it tends to compress the plasma. In the high current discharge, the plasma will be detached from the wall of the discharge vessel and will form a very thin filament along the axis of the vessel (or tube).

pinch instability a small perturbation can cause a slight bending of the plasma column. The field lines thereby are crowded inside the bend and separated outside. As a result, the magnetic flux density is increased on the inside of the bend and causes the column to bend still more, giving rise to the **pinch instability**.

pitch angle the pitch angle θ refers to a particle in a magnetic mirror. It is defined by tan $\theta = v\perp/v_{11}$.

plasma one can define plasma in any one of the following ways:

(i) a fluid containing sufficiently numerous free charged particles causing its dynamical character to be dominated by electromagnetic fields
(ii) a quasi-neutral gas of charged and neutral particles
(iii) a fully ionized gas of low density
(iv) a coulomb gas
(v) a fourth state of matter resembling gaseous state at high temperature.
(vi) a system of charged and neutral particles satisfying certain criteria, e.g. $L >> \lambda p >> n_e^{-1/3}$ and $\nu < \omega p$.

plasma chemistry plasma chemistry makes extensive use of the fact that chemical reactions in the gas phase or at surfaces can occur at low process temperature when the atoms are excited by a gas discharge.

plasma confinement retention of plasma within a region, including the physics of heat and particle losses from the plasma. See transport.

plasma crystal regular arrangement of highly charged particles in dusty plasmas. Under the influence of their mutual repulsion these particles form a Coulomb crystal.

plasma current the current flowing toroidally through the plasma in a TOKAMAK. See poloidal field.

plasma etching a dry-etch technique in which reactive atoms or ions (like oxygen, chlorine, fluorine) are generated in a gas-discharge (mostly rf-operated). Reactive ion etching makes use of the fact that ions are accelerated in the boundary layer between plasma and substrate with high directivity and are thus able to create deep grooves with steep walls.

plasma frequency it is given by

$$\omega_p = \left(\frac{4\pi n_0 e^2}{m}\right)^{\frac{1}{2}}$$, where the symbols have usual significance.

plasma geometry the position and shape of the plasma cross-section. See major radius, minor radius, aspect ratio, elongation, triangularity.

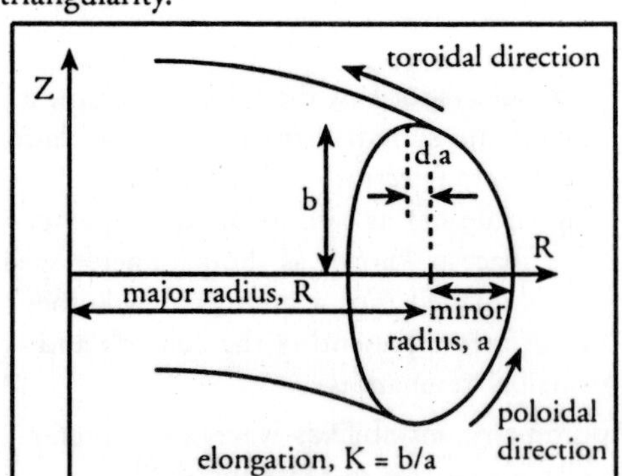

plasma gun it is one type of plasma accelerator, which was first, developed by Marshall for use in thermonuclear fusion research and has since been adapted and tested for rocket propulsion. Plasma is driven out of a hole at one end of the gun.

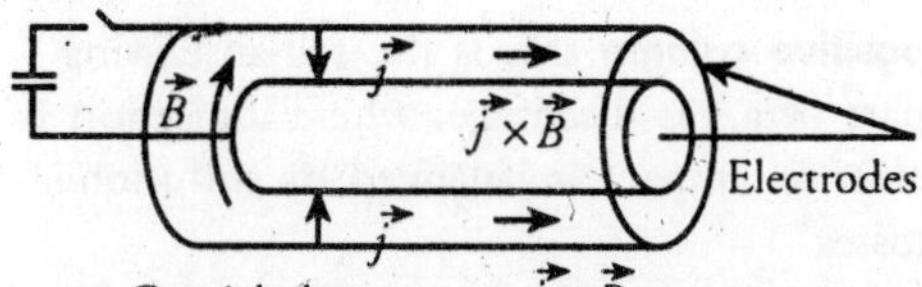

Coaxial plasma gun $\vec{j}\times\vec{B}$

The driving mechanism is the direct $\vec{j}\times\vec{B}$ force (Fig). This force is introduced in the gun by the co-axial electrodes which allow the passage of a radial current. As the plasma moves along the gun, this radial current move with it, thus propagating the action of the Lorentz force.

plasma lens a plasma lens has unusual optical properties, since the index of refraction is less than unity.

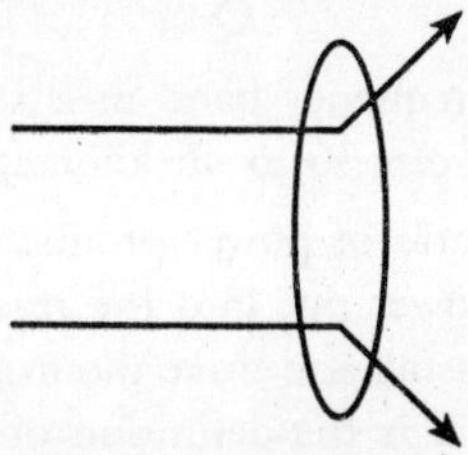

plasma propulsion by applying to the plasma both an electric field and magnetic field, perpendicular to each other, one can produce a drift velocity $(\vec{E}\times\vec{B})/B^2$, having magnitude E/B; thus one can obtain mechanical energy from electrical energy. Thus by making E/B sufficiently large, one can obtain a plasma beam with a high enough velocity to act as an ordinary rocket-propulsion system.

plasma radiation radiation (i.e. energy loss via electromagnetic waves) from a plasma is due to a number of processes. See bremsstrahlung, line radiation, recombination radiation, synchrotron radiation, Impurity radiation.

plasma rotation bulk rotation of the plasma in the toroidal or poloidal direction. Neutral beam injection can cause plasma rotation in the toroidal direction at velocities of typically 100 km/s.

plasma sheath assume that the potential ϕ inside the plasma is zero. When ions and electrons of the plasma hit the wall of the container, they recombine and are lost. Since electrons have much higher thermal velocity than ions, they are lost faster and leave the plasma with a net positive charge. The plasma will then have a positive potential with respect to the wall, that is, the wall potential ϕ_w, is negative. Debye shielding confines this variation in ϕ to a layer of a few Debye lengths in thickness. This layer must exist on all cold walls, with which the plasma is in contact and is known as a sheath. The sheath is equivalent to a potential barrier to the electrons. The height of the barrier self-adjusts, so that electron flux going over the barrier to the wall is just equal to the flux of ions reaching the wall.

plasma simulation with today's computers the collective behavior of plasmas can be investigated by solving the equation of motion of individual particles. Such simulations give detailed insight into microscopic processes that, in real plasmas, are inaccessible by conventional measuring techniques.

plasma technology common name for a number of industrial applications of plasma, such as: etching of semiconductor chips, deposition of silicon for solar cell production, deposition of silicon dioxide for passivation of surfaces, activation of surfaces, melting and welding with plasma arcs as well as plasma chemistry.

polarimetry measurement of the rotation of the plane of polarisation of light; used to measure the local magnetic field and, thus, the safety factor. See Faraday rotation.

polarisability it is given by $\vec{P}(\vec{r}, t) = \alpha\vec{E}$, where $\vec{P}$ = polarisability of the plasma, = induced electric dipole moment per unit volume and $\vec{E}$ = electric field produced due to charge imbalance.

poloidal beta the poloidal component of the plasma beta—a measure of plasma confinement by currents flowing poloidally in the plasma. In a TOKAMAK, the poloidal beta is equal to 1 if no poloidal currents are flowing. In this case, the plasma is confined entirely by the toroidal plasma current crossed with the poloidal magnetic field it induces. If the poloidal beta is zero, the plasma current lies entirely parallel to the magnetic field, and there is no magnetic confinement—a region of plasma where this applies is known as a force-free region. If the poloidal beta is greater than one the plasma is confined mostly by the poloidal current crossed with the toroidal field, and the toroidal current merely serves the purpose of centering the plasma. This latter configuration is sometimes referred to as a high beta device.

poloidal field component of the magnetic field parallel to the minor circumference. The poloidal field is essential for confinement and, in a TOKAMAK, is generated by the plasma current (c.f. stellarator); this is in contrast to the larger toroidal field, which is generated externally.

poloidal magnetic field this field is created by sending a current along the toroidal direction. The object is to avoid closure of magnetic lines of force once or several times around the torus, so as to generate a small rotational transform, as explained below:

In an ordinary torus, the lines of induction are closed. If the lines of force or the torus is twisted in some direction, a set of lines of force, after a complete revolution inside the torus, will be transformed into another set of lines of force. In the next revolution, this (transformed) set of lines of force will be transformed into a third set. If, after each revolution, rotation takes place in the same direction, it is known as the rotational transform.

ponderomotive force light waves exert a small radiation pressure. But when high power microwaves or laser beams are used to heat or confine plasmas, the radiation pressure can be several thousand atmospheres. When applied to a plasma, this force is coupled to the particles in some way and is called the ponderomotive force. Many non-linear phenomena can be described in terms of the ponderomotive force.

positive column this is the self-sustaining part of a gas discharge, where ionisation by electron impact is balanced by diffusion losses.

power plant a fusion power station, producing economically-feasible amounts of electricity from fusion reactions (See blanket). The term "reactor" is sometimes used for such a device, or for any fusion device in which the plasma is ignited.

power threshold the L-H transition (and also improved performance regimes related to reverse shear) occurs when the power exceeds a certain threshold value—the power threshold.

Q

Q-band a frequency band used in radar extending from 36 to 46 kilomegacycles.

Q factors ratio of power produced by fusion to power put into the reactor to heat the plasma and drive the magnetic fields. Q = 1 is the definition of scientific breakeven, where power cut = power in economical fusion will require Q significantly greater than 1.

Q limit the safety factor at the edge of the TOKAMAK plasma must be greater than 2 in order to avoid disruptions. (This limit becomes more restrictive at tight aspect ratio.) See Hugill diagram.

Q-machine a Q-machine produces quiescent plasma by thermal ionization of CS or k atoms impinging on hot tungsten plates. Diamagnetic drifts were first measured in Q-machines.

Q switched laser a device for producing very short intense laser pulses by enhancing the storage and dumping of electronic energy in and out of the lasing medium.

Q switch optical switch used to rapidly change the Q of an optical resonator, utilized in the optical resonator of a laser to prevent lasing action until a high level of gain and energy storage are achieved in the laser medium. When the switch is triggered and rapidly increases the Q of the cavity, a stronger laser pulse is generated.

quadrature

1. an elongation of 90 degrees usually specified as east or west in accordance with the direction of the body from the sun. The moon is at quadrature at first and last quarters.
2. the situation of two periodic quantities differing by a quarter of a cycle.

quadrupoles a linear accelerator having four longitudinal vanes in its resonating cavity, which are shaped to create RF electric fields that simultaneously accelerate, bunch and focus the charged particle beam.

qualitative analysis an analysis in which some or all of the components of a sample are identified.

quality control an aggregate of functions designed to ensure adequate quality in manufactured product by initial critical study of engineering design, materials, processes, equipment, and workmanship followed by periodic inspection and analysis. used by reliability control.

quality factor a measure of the sharpness of resonance or frequency selectivity of a resonant vibratory system having a single degree of freedom, either mechanical or electrical. In a mechanical system, this quantity is very nearly equal to one-half the reciprocal of the damping ratio.

quantization the process of converting from continuous values of information to a finite number of discrete values.

quantities the values that mark frequency distribution interval boundaries that are determined by arranging a set of N observations in order of magnitude and marking off equal parts (N/P) of the total population.

quantize to limit the possible values of (e.g. quantity) to a discrete set of values by quantum mechanical rules.

quadrature amplitude modulation amplitude modulation in which the signal is 900 out of phase with the carrier which it modulates.

quantum chromodynamics a gauge theory describing the interaction between quarks and gluons. Used for colour (particle physics) and QCD.

quantum dots small (30nm to 1 micron) metal or semiconductor structures that hold a discrete number of electrons; the number of electrons can be altered by modifying the electrostatic environment of the dot.

quantum efficiency a measure of the efficiency of conversion or utilization of light or some other form of energy.

quantum electronics the branch of electronics that essentially deals with lasers and laser devices which require quantum theory for their exact description.

quantum wells effective potential wells created by a minimum in the conduction band or a maximum in the valance band that arises when a smaller band gap semiconductor is sandwiched between a larger band-gap semiconductor.

quark proton model a theoretical model which summarizes our understanding of how protons and neutrons are made up of the fundamental sub-particles called quarks.

quasars a specific type of quasi-stellar source. A stellar-appearing object of very large redshift that is a strong source of radio waves; presumed to be extragalactic and highly luminous.

quartz crystalline silica, an important rock-forming mineral.

quasi-neutrality a plasma is quasi-neutral, that is, almost neutral, which means that there cannot be any significant excess positive or negative charge accumulated at any point in it. This is expressed as:

$n_i \approx n_e >> | n_i - n_e |$,

n_e, n_i being the number density of electrons and ions, respectively. Consider the situation if the plasma deviates from charge neutrality even slightly. For, let us consider a plasma sphere of 0.01m radius. let $n_e \approx 10^{17} m^{-3}$. if n_i exceeds n_e by even 1%, the electric field generated

$$|\vec{E}| = \frac{4}{3}\pi r^3 \frac{(n_i - n_e)}{4\pi\varepsilon_0} \frac{e}{r^2} = 60{,}000 \frac{V}{m}$$ which is

too large to be maintained in a plasma, as the particles are free to move and, hence, quickly neutralize regions of excess charge by transfer of electrons, of which the acceleration is $eE/m=10^{16}m/sec^2$. This is the meaning of quasi-neutrality.

quasi-Lagrangian co-ordinates a system of mixed Eulerian and Lagrangian co-ordinates. At least one co-ordinate of each fluid parcel must therefore be unvarying with time.

Que frequencies in spectral analysis, the frequency of periodic ripples in a spectrum of a signal that contains echoes. They are expressed in c/s or in hertz.

quenching (atomic physics) phenomena in which very strong electric fields cause the orbit of an electron or atom to precess rapidly so that the average magnetic moment associated with its orbit angular momentum is reduced to zero.

quenching (cooling) rapid cooling as in metallurgy. Used for plasma quenching.

query languages command languages used to search and retrieve information.

quiet sun the Sun when it is free from unusual radio wave or thermal radiation such as that associated with sunspots.

quinoxalines a group of heterocyclic compounds consisting of a benzene ring condensed with a diazine ring.

R

radiation belts (Van Allen) the magnetic traps created by nature for electrons and ions in the magnetic field of the earth are known as the Van Allen radiation belts. They are composed of very highly energetic electrons and ions. The origin of these particles is not known. Some mechanisms connected with the cosmic rays incident on the Earth from outer space and plasma ejected from the Sun during solar flare are probably responsible for the creation of these particles.

recombination radiation radiation produced when an (impurity) ion recombines with an electron.

recombination, coefficient of when an ion and an electron moving at relatively low velocities happen to collide, a neutral atom is formed. This process is called recombination. It is a loss process and can be represented as a negative source term in the equation of continuity, proportional to $n_e n_i = n^2$. Thus, $\partial n/\partial t = -\alpha n^2$, the constant of proportionality α is called the coefficient of recombination.

recycling in most TOKAMAKs the pulse length is much longer than the particle confinement time. Thus on average each plasma ion goes to the wall or limiter and returns to the plasma many times during the length of the discharge. This process is called recycling.

resistive ballooning modes a class of ballooning mode which would be stable in the absence of resistivity, but can be unstable in its presence. Related to, but topologically different from, tearing modes.

resistive instability instability due to diffusion and rearrangement of magnetic field lines. When the plasma resistivity is small, these instabilities have a slow growth rate.

resistivity the tendency to resist the flow of electric current, thereby dissipating energy. Plasmas are very good conductors of electric current, so that as a first approximation their resistivity can often be neglected. In this case, "ideal" magnetohydrodynamics may be applied.

resonances and cut-offs (definitions) resonance is defined as $V_{ph} = 0$ or $\varepsilon = \infty$ and cut-off is defined as $V_{ph} = \infty$ and $\varepsilon = 0$. In the case of cut-off, waves are generally reflected, that is why cut-off point is alternately known as reflection point.

resonant angle it is the angle between $\vec{K}$ and $\vec{B}$. the angle θ for which $V_{ph} = 0$ i.e. resonance occurs, is known as the resonant angle.

resonant ions/electrons resonance occurs when one of the characteristic frequencies of particle motion in the plasma (for example, the cyclotron frequency) matches the frequency of some applied perturbation (for example, an RF wave).

resonant magnetic perturbation an externally applied magnetic perturbation matched to the spatial structure (and optionally the frequency and phase) of an instability.

reverse field pinch a toroidal magnetic confinement device in which the poloidal and toroidal fields are of comparable magnitude. To maintain stability the toroidal field reverses close to the edge of the plasma when a critical plasma current is exceeded.

reverse (magnetic) shear in a TOKAMAK the current density is usually greatest at the magnetic axis, in which case the safety factor rises from the centre to the edge of the plasma. Using non-inductive current drive and/or the bootstrap current the current density can be made to be greatest away from the centre. In this "reverse shear" case, the safety factor has a minimum away from the plasma centre. Using reverse or low shear ("optimised shear") some TOKAMAKS, notably DIII-D and TFTR in the US and more recently JT-60U in Japan and JET, have shown greatly improved plasma performance. Reverse shear is an attractive option for advanced TOKAMAK scenarios.

Reynolds numbers if this number exceeds 0.1, laminar flow is stable for all Reynolds numbers.

Rogowsky coil in order to measure the current density directly a Rogowsky loop (fig) is used; the plasma current generates a field that produces a current in the Rogowsky coil, which is wound around the plasma.

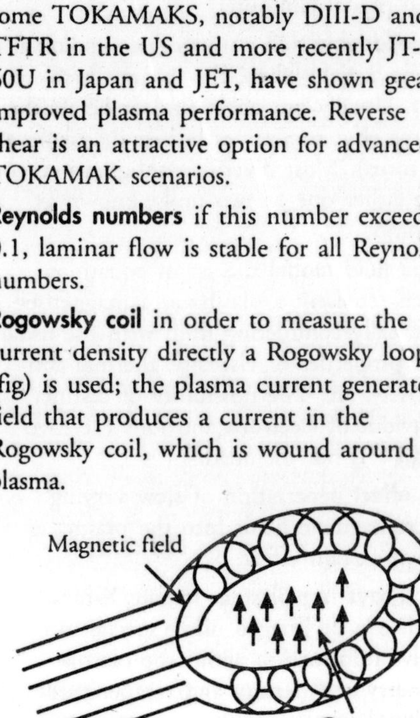

rotational transform (see poloidal magnetic field).

runaway electrons even for very weak electric fields, some very fast electrons never reach a steady balance between acceleration by the field and deceleration by collisions. These electrons which are accelerated by the field at a rate greater than they are slowed down by collisions are known as **runaway electrons.**

S

safety factor, q number of turns the helical magnetic field lines in a TOKAMAK make round the major circumference for each turn round the minor circumference. Denoted by the symbol *q*. There is no connection with the ordinary sense of the word "safety"

sawtooth a cyclically recurring instability which affects the central region of TOKAMAK discharges. The temperature periodically falls abruptly, then slowly recovers. The jagged trace produced by plotting temperature against time gives the instability its name

scaling laws empirical or theoretical expressions for how various plasma phenomena (e.g. confinement, power threshold, etc.) vary with the TOKAMAK conditions using a range of free parameters to be fixed by "best fits" of the scaling law to TOKAMAK data. They are particularly useful for predicting the performance of future TOKAMAKS.

second stability a second region of stability against ballooning modes, which occurs at high plasma pressure gradient. It can be obtained either by adjusting the required pressure and q profiles in a TOKAMAK of usual cross-section, or by shaping the plasma boundary to a "bean" shape. Second stable plasmas have a high poloidal beta, which leads to a high bootstrap current. This has potential cost savings for a fusion power plant as a less powerful current drive system would be required.

shear refers to either the variation of plasma flow (flow shear) or safety factor (magnetic shear) with minor radius. If the type of shear is not specified, it usually means magnetic shear.

shield major component of a fusion power plant, which reduces the neutron flux reaching the TF coils and beyond. This minimises the radiological impact of the neutrons, and their heating of the coils which, if superconducting, need to remain at liquid helium temperatures.

shock heating it is a method in which the plasma is compressed and heated by a shock wave.

shock waves supersonic flow in a nonconducting fluid can give rise to shock waves, i.e. surfaces at which the pressure, density and stream velocity change abruptly over a short distance. This distance is determined by the viscosity and thermal conductivity of the fluid. In a gas, the thickness of the shock front is of the order of a mean free path.
In plasma there can also be magnetohydrodynamic shock waves into which magnetic pressure enters together with the ordinary kinetic pressure. The role of the sound speed is played by the Alfvén speed and the role of the mean free path by the cyclotron radius or Debye length
the structure of the shock front in plasma is complicated and has an oscillatory character.

sigma (σ) effect perturbation of the electric field between the electrodes by the space charge and consequent production of a spark. It is with the help of this effect that Raëthu and Meek established their famous streamer theory.

Simon diffusion consider plasma in a cylindrical tube with an axial magnetic field directed along the tube. The plasma particles diffuse across the magnetic field into the wall. The diffusion rate depends strongly on the material used to block off the ends of the tube. If insulated ends are used, the charges that arise in the plasma are not neutralized. In this case, the diffusion is ambipolar. The diffusion rate will be twice the rate of electron diffusion and much less than for ions diffusing across the field alone. However, if the insulated ends are replaced by metal ends, the character of the diffusion is completely changed. In this case, the diffusion will cease to ambipolar and its rate will then be determined by the larger of the two particle diffusion coefficients. This is called Simon diffusion.

stimulation of plasma each particle in a plasma, if it is charged, set up its own electrostatic, magnetic and radiation fields, which react on all the other particles. In principle, we could compute the net effects of each particle on every other particle, as well as include the effects of externally imposed $\vec{E}$, $\vec{B}$ and $\vec{g}$. Still follow the trajectories by solving the equation of motion simultaneously. However we would need a computer that could follow the billions and billions of particles making up a plasma all at once. Such computers are not available, but substantial progress in the realm of plasma simulation has already been achieved on computers. This simulation involves the simultaneous solution of the equations of motion of a "super-particle", which is often defined as a rigid "clouds of particles" moving through other such "clouds". Up to about 100,000 or 500,000 of such interacting superparticles can be moved simultaneously on a high-speed digital computer.

single/double null points of zero poloidal magnetic field where the separatrix crosses itself. Usually put above and/or below the plasma (but sometimes put in the plane of the torus). Most divertor configurations have either one or two nulls, known as X-points.

single fluid model the set of equations which represent a plasma as a magnetised, electrically conducting fluid with the usual fluid properties of viscosity, thermal conductivity, etc. The possibility of distinct behaviour of electrons and ions (i.e. two "fluids") is not included.

skin effect penetration of slow varying electromagnetic fields into the plasma is known as skin effect.

soft X-rays hot plasmas radiate X-rays, which can be imaged using special cameras to give information about the plasma geometry and density and temperature profiles.

solar corona over the Sun's surface, there is a low density high temperature region called the solar corona. It is a rather powerful source of ultraviolet radiation. The Sun emits plasma from its surface.

solar flare it is a very complex event regularly occurring in the solar system and releasing enormous amount of energy ($\approx 10^{25}$ joule) per event. It causes radiative emission of X-rays, ultraviolet-, optical- and radio-frequencies as well as acceleration of charged particles to relativistic energies. It originates from rapid compression of plasma by a magnetic field in the deep space. The flare expelled from the Sun into space as a plasma stream, is captured by this magnetic field in various kinds of magnetic traps. After a solar flare, magnetic storms and auroras are observed on the Earth and radio communication is disturbed. Shock waves and artificial satellites have confirmed that solar flares cause disturbance outside the earth's atmosphere by plasma streams and by shock waves propagating in the interplanetary plasma.

solar wind a plasma wind known as the **solar wind** is ejected from the Sun and impinges on the upper atmosphere of the Earth with significant effects. This stream has no magnetic field embedded in it. According to the field-freezing concept, an external magnetic field should be unable to penetrate such a plasma. There is a random interplanetary magnetic field in the solar system. The plasma stream arriving from the Sun displaces this external magnetic field. It might be said that the "plasma broom" sweeps the interplanetary magnetic field from the vicinity of the Sun.
Solar wind consist of a continuous stream of charged particles (protons and electrons) having all the usual plasma properties; it compresses the Earth's magnetic field B until the wind pressure is equal to $B^2/2\mu_0$.

solitons solitons (or solitary waves) are electrons or electromagnetic wave packets (or pulses). They are a potential and density disturbances propagating with a certain velocity. Solitons have been observed in both laboratory experiments and computer solutions. They are remarkably steady. Two solitons propagating with different velocities interact when they overlap non-linearly. Then they part company each one preserving its original shape and velocity. A non-linear wave equation known as the Kortewig de Vries or kdV equation describes the evolution of the solitary wave.

spectroscopy the detection and analysis of the spectrum of radiation emitted by a plasma. This can yield information about temperatures, impurities, rotation, etc. can use different parts of the electromagnetic spectrum (infra-red, visible, VUV, XUV, X-rays, etc.)

spherical torus, spherical TOKAMAK a very low aspect ratio torus approximating to a sphere (although topologically remaining a torus). Very low aspect ratio TOKAMAKs are often called spherical TOKAMAKs. See START, MAST, NSTX

spheromak spherical plasma in which comparable toroidal and poloidal currents flow. The toroidal current is not driven by transformer action.

stability theory the theory of how small perturbations to a system evolve in time. Spontaneous growth is instability. Instabilities can saturate at some small amplitude, in which case they degrade confinement, or grow uncontrollably, in which case confinement is lost (e.g. a disruption occurs).

START Small Tight Aspect Ratio Tokamak at Culham, the pioneering device of its type. START has now been superseded by MAST; the machine itself has been transferred to ENEA research laboratory in Frascati, Italy, where it will be used for further plasma experiments. See internal reconnection event, beta. *START Home page.*

stellar spectroscopy since Fraunhofer's investigations of the dark line in the spectrum of the Sun (1813), we have gained detailed knowledge about chemical composition, temperature and density of the stars from the wavelength and shape of spectral lines.

stellarator it is possible to have a trap in which the plasma is isolated from the walls by a strong external longitudinal field and the intrinsic field only plays a subsidiary role. Traps with currents that flow along

the magnetic field are subject to various hydromagnetic and kinetic instabilities. It is useful to device a trap in which the helical field is established by currents that flow in external conductors; This type of trap is called a stellarator. The name "Stellarator" comes from the word "Stellar" (star-like). This name is used because it is hoped to achieve temperature of the same order as those found in the regions of stars in which thermonuclear reactions occur.
In its original form, the stellarator was a torus twisted into a figure of eight.

stuart number it is a dimensionless number which determines the character of the flow in the limiting case of small magnetic field.

super Alfvénic velocity a velocity greater than the Alfvén velocity, which is a typical velocity associated with magnetohydrodynamic behaviour. In a TOKAMAK, only energetic particles have super Alfvénic velocities; because they satisfy this condition, they may resonantly transfer their energy to magnetohydrodynamic modes, which may grow as a result (e.g. TAE modes).

superthermal radiation electromagnetic radiation produced by energetic particles, as opposed to thermal particles. Typically in the cyclotron frequency range.

synchrotron radiation radiation emitted by hot plasma in a magnetic field is known as magneto-bremsstrahlung or synchrotron radiation. In a magnetic field, the thermal motion of the plasma particles is a combination of unrestrained motion along the magnetic field lines and cyclotron gyration around them. The gyration of the electrons around the magnetic field lines leads to magnetic radiation. This form of radiation only occurs in very hot plasmas which contains relativistic electrons. Relativistic electrons are not produced in a cyclotron, but in another kind of accelerator known as a synchrotron. Therefore, the magneto-bremsstrahlung radiation from relativistic electrons is called synchrotron radiation. This form of plasma radiation is very important in astrophysics. A strong source of synchrotron radiation is the Crab Nebula.

T

TAE modes Toroidal Alfvén Eigenmodes. similar to Alfvén gap modes.

target in laser fusion the deuterium-tritium fuel is contained in glass micro-balloons of 0.2-0.5 mm diameter.

tearing magnetic islands the disturbance (caused by a tearing mode) which alters the topology of the confining magnetic field and permits transfer of heat across the affected region.

tearing mode a class of resistive MHD instability which has been predicted theoretically in TOKAMAKs. Positively identified in experiments. See tearing magnetic islands, neo-classical tearing modes.

temperature from the thermodynamical viewpoint, temperature characterizes the energy distribution of the particles (in the case of a plasma, the electrons and ions). The energy distribution is Maxwellian. The thermal energy is expressed as $\frac{1}{2}m\bar{c}^2 = \frac{3}{2}KT$, where the symbols have their usual significance.

temperature, longitudinal and transverse in a plasma in a magnetic field, the velocities parallel and perpendicular to the field can be assumed to be Maxwellian. Under these conditions, we can say that the plasma has two temperatures : one, corresponding to motion along the field, called longitudinal temperature and denoted by T_{11} and the other, corresponding to motion at right angles to the field called transverse temperature and denoted by $T\perp$. if $T_{11} \neq T\perp$, the temperature is anisotropic. Likewise $p_{11} \neq p\perp$, that is, pressure is also anisotropic.

temperature pedestal in H-mode there is a region of steep temperature gradient at the plasma edge. The temperature at the top of this steep gradient region is the temperature pedestal.

TF coils the coils that produce the toroidal field in a TOKAMAK.

TFTR the TOKAMAK Fusion Test Reactor at Princeton was the largest US device, and performed a major campaign using deuterium and tritium fuel between 1993 and 1997. TFTR had a relatively high magnetic field of 5 T and a circular cross-section. See reverse shear. *TFTR Home Page.*

thermal cycling successive heating and cooling of materials (by transformer action, for example) that can lead to cracks rupture, etc., particularly at boundaries between materials that expand at different rates.

thermal particles as a result of collisional energy exchange, the energy of most plasma particles falls within a distribution which can be described by a single temperature (typically 1 to 30 keV for tokamaks). These are the thermal particles, as distinct from energetic particles which lie outside the thermal equilibrium.

thermionic converter the thermionic converter is a device where cesium plasma is used between two electrodes, one heated to emit electrons and the other cooled. Because of the cesium plasma, very high current are produced and a significant portion of the thermal energy applied to the cathode is extracted as electric energy.

theta pinch (See linear pinch)

Thomson scattering diagnostic diagnostic to measure temperature and density by detecting laser light scattered and Doppler shifted by the plasma electrons.

TOKAMAK it is a device to achieve fusion in a toroidal chamber subject to a magnetic field. The word is a brief Russian name of which the full meaning is "Toroidal Magnetic Chamber". Russian physicist L.A. Artsimovich first evolved this device in 1968. This is why he is called the father of TOKAMAK research. It is at present considered to be the most promising one for controlled thermonuclear fusion.

The most successful device yet found for magnetic confinement of plasma. Its magnetic field is made up from helical lines of force on toroidal surfaces, and is generated both by external field coils and by the current in the plasma.

TOKAMAK operating boundaries the set of plasma parameters (e.g. density, current, pressure), beyond which it is impossible to operate a TOKAMAK. Careful use of plasma cross-sectional shapes and current profiles can increase the operating regime. See Hugill diagram.

toroidal field the largest component of the magnetic field of a TOKAMAK, produced by coils external to the plasma, which is in the toroidal direction.

toroidal stability stability taking account of effects due to the toroidal geometry. These are sometimes neglected to identify possible instabilities, but must usually are included for accurate predictions of stability boundaries

toroidal trap a magnetic field keeps the plasma from drifting to the side walls, but has no effect on motion towards the ends of a tube. A simple way of producing a magnetic trap is to make an endless tube. We simply bend the tube into a circle. A body similar to a doughnut has a special geometric name, the torus. A magnetic trap of this form is referred to as toroidal. However, plasma confinement in a toroidal trap is complicated by the fact that the magnetic field cannot be homogeneous in such a trap. To confine a plasma in a torus, it is necessary to use a more complicated magnetic field.

torsatron a stellarator configuration with continuous helical coils. modular designs in which the continuous coils are replaced by a number of discrete coils producing a similar field are also possible.

Townsend criterion of breakdown it is given by $\gamma(e^{\alpha d} - 1) = 1$ where α = Townsend's first coefficient, defined as the number of ion-pairs on the average that can be created by an electron in traversing a distance of 0.01m and γ = Townsend's Second coefficient, defined as the number of secondary electrons ejected from the cathode by an ion (produced in the volume by an electron) reaching it, and d = distance between electrodes.

trapped particles the outside (larger major radius portion) of a TOKAMAK plasma has a lower magnetic field than the inside. Particles with low velocity parallel to the magnetic field may not have sufficient energy to enter the higher field (inside) region and become trapped on the outside. They are not free to circulate toroidally but instead bounce back and forth, performing so-called banana orbits. (See trapping below for an alternative description.)

trapping particle trapping occurs over the central region of most TOKAMAKS, where the collisional mean free path is longer than the distance a particle must travel along a field line to get from top to bottom of the torus (the connection length). It results from the conservation of total particle energy and magnetic moment. If a particle's motion is nearly perpendicular to the magnetic field, it can become trapped in an orbit that does not complete a revolution in either the toroidal or poloidal direction.
Since this orbit has a finite width due to the particle's drift across the magnetic field it is known as a banana orbit. Particles not trapped in this way are called passing particles.

tritium isotope of hydrogen with a nucleus consisting of one proton and two neutrons. Tritium is radioactive with a half-life of 12.3 days.

turbulence it is known in ordinary hydrodynamics that for large velocity, the laminar flow becomes unstable and turbulence occurs. Viscosity improves stability, so that Reynolds number can be used as a rough measure for the turbulence limit. Turbulence impedes the flow considerably. Now the presence of a magnetic field helps viscosity in preventing turbulence and, therefore, the critical Reynolds number. Accordingly, for large velocity the flow will be increased by the presence of a magnetic field, although small velocity (laminar flow) is decreased by it.

turning point it refers to a particle trapped between two magnetic mirrors. It bounces between them and, therefore, has a periodic motion at the "bounce frequency". However, since the guiding centre drifts across field lines, the motion is not exactly periodic and the constant of the motion becomes an adiabatic invariant. This is called **longitudinal invariant** J and is defined for a half-cycle between the two turning points

$J = \int_a^b v_{11} ds$

two-fluid model and multi-fluid model the extended set of equations which represent a plasma as inter-penetrating and interacting fluids of electrons and ions, impurity ions etc. (c.f. single fluid model).

U

ultraviolet radiation this radiation from the sun produces the relatively high degree of ionization in the ionosphere during the daylight hours and when this radiation source is "turned off" at sunset, the electron number density falls off gradually to a much lower equilibrium value.

upper hybrid frequency this is defined by $w^2_{uh} = w^2_{pe} + w^2_{ce}$ (only the electrons vibrate). the lower hybrid frequency is defined by $w^2_{lh} = w_{ci}\, w_{ce}$ (both electrons and ions vibrate and their velocities along $\vec{K}$ are approximately equal).

V

Van Allen radiation belts the magnetic traps created by nature for electrons and ions in the magnetic field of the Earth are known as the Van Allen Radiation Belts. Its existence was first established in 1958 by Van Allen and his co-workers during launching of artificial satellites and rockets. these workers found that cosmic radiation detectors on board the satellite, Explorer I, stopped functioning at distances about 700 km from the earth and beyond; however, when the orbit of the satellite brought it within this distance, the counters behaved normally again. They concluded that this behaviour was consistent with the theory that above about 700 km the counters were being choked by an anomalously high radiation intensity. Laboratory test showed that this would happen if the intensity of radiation exceeded a value about 10^4 times greater than that due to cosmic radiation. The hypothesis that the satellite passed through regions of high radiation intensity was tested by further experiments on board other satellites in the explorer series. By putting counters on board lunar probes, regions of high radiation intensity were detected at great distances from the earth. This and later work helped to establish the morphology of the radiation belts.

The particles in the belts are mostly electron and protons. The protons in the inner belt have energies 40 Mev and above, deutrons and tritons are also present, but only about 1% of the number of protons. The energies of protons in the outer belt are much lower (≈ a few Mev); the picture is not quite so simple as that of the figure, since it is now known that there are several low energy proton belts around the earth and even these overlap. The outer belt of electrons is peaked at about 30,000 km and so it is now customary to consider proton and electron belts separately. Electrons in the outer belt have energies typically greater than 40kev.

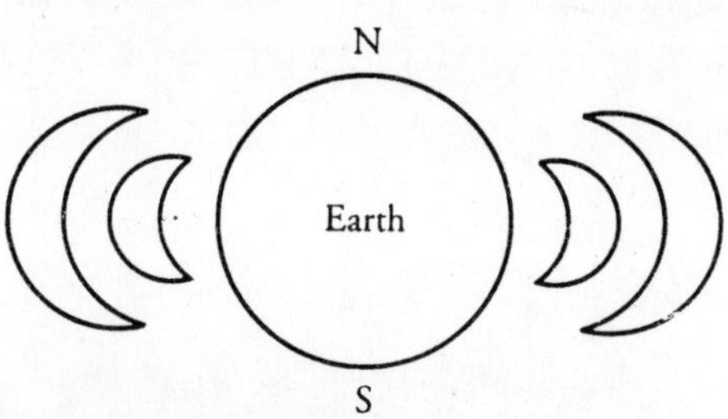

vertical displacement event a plasma instability during which the whole plasma moves up (or down) away from its equilibrium position. High elongation plasmas are more prone to this motion due to the more strongly shaped magnetic field required to produce them. Unless it is controlled by a feedback system the plasma will be rapidly lost as it collides with the vessel, causing its current to flow through the wall and other components. See halo current.

viscous dissipation if the plasma moves as a whole, then collisions will manifest themselves as an internal friction; this is viscous dissipation. in a dense plasma, this dissipation mechanism is dominant. In very tenous plasmas, the damping is independent of collisions.

Vlasov equation it is collisionless Boltzmann equation (See Boltzmann-Vlasov equation.

VUV the "Vacuum Ultra Violet" range of the electromagnetic spectrum.

W

warm plasma refuelling fuelling of plasma using medium energy particles of particle clusters.

wave vector it is the same as propagation vector and its magnitude is given by $\vec{k} = 2\pi/\lambda$.

Wendelstein 7-X (W7-X) a large advanced stellarator under construction in Germany to follow its present W7-AS device, based on the Helias configuration. *W7-X Home Page.*

whistlers whistlers are produced by electrical discharges in the atmosphere (thunderstorm), which excite electromagnetic waves in the radio frequency range. These waves propagate mainly along the field lines of the Earth's magnetic field, so that they re-enter the atmosphere at a point on the opposite hemisphere. since these frequencies obey $w_{ci} << w << w_{ce}$, they propagate as left circular transverse waves with a refractive index approximately given by

$$n(w) = \frac{w_p}{\sqrt{(w_{ce}w)}}$$; accordingly the group

velocity is given by

$$\frac{dw}{dk} = \frac{2c}{w_p}\sqrt{w_{ce}w}$$ and, therefore, increases

with frequency. Consequently, the highest frequencies arrive first, the lower frequencies later. This is the reason why whistlers are received as short whistles, whose pitch is high at the beginning and subsequently drops. The same whistler may be received several times, because it is reflected back and forth along the field lines.

work function the work function is the energy needed by an electron to leave a metal surface.

X

X-rays (soft X-rays) hot plasmas radiate X-rays which can be imaged using special cameras to obtain information about the plasma geometryand density and temperature profiles.

XUV the "Extreme Ultra Violet" range of the electromagnetic spectrum. shorter wavelengths than VUV.

Z

Zeeman splitting a spectroscopic method based on this splitting of spectral lines can be used to measure a magnetic field.

Z-effective the effective charge of the plasma. Z_{eff} is a measure of how free from impurity ions the plasma is (its value rises with the square of the ion charge). For a pure D-T plasma, Z_{eff} = 1.0. In a fusion power plant its value will have to be 2.0 or less.

Z-pinch (See linear pinch).